Seahog's Deep Sea Adventure

Nama

CONTENTS

1 INTRODUCTION

The development of the Seahog started in 2010. It is a third generation tethered underwater Remotely-Operated Vehicle currently in development by the Robotics and Agents Research Lab (RARL) within the Department of Mechanical Engineering at the University of Cape Town. The Seahog was established as an overhaul of two previous ROVs developed by various students within the same lab. The reason for building an in house ROV was to develop a cheap alternative to commercial offerings to specifically meet the needs described by the marine biology group at the University of Cape Town. They wanted an ROV mainly for marine life observation with a possibility of small sample collection. The Seahog was started by two master's degree students after it was deemed unfeasible to improve upon the Challenger I or Robin (the names given to the two previous ROVs). Development of the open-frame ROV Seahog was then commenced. This design was perceived to be a better design compared with the two previous ROVs' which were torpedo-shaped. This design choice offered more in terms of adaptability and maintenance as each functional feature was contained in its own module separate from other modules. For example, the Inertial Measurement Unit (IMU) is housed in its own module, so in case of a navigation feedback breakdown, debugging would be easier to perform.

Table 1 shows some of the specifications or characteristics of each of the three ROVs developed by the RARL lab over the years.

Table 1: Some specifications of the ROVs developed by the RARL lab

	Challenger I	*Robin*	*Seahog*
ROV generation (development year)	1st (2009)	2nd (2010)	3rd (2010)
Dry weight	52 kg	52 kg	82 kg
Size (L × W × H)	0.72 × 0.535 × 0.722 m	0.741 × 0.543 × 0.422 m	0.87 × 0.66 × 0.5 m
Depth rating	60 m	160 m	300 m
Maximum thrust (per thruster)	12 N	18 N	60 N
Number of thrusters	4 (2 horizontal, 2 vertical)	3 (2 horizontal, 1 vertical)	5 (4 horizontal, 1 vertical)
Navigation capacity	Heading/depth control	Heading/depth control	Heading/depth control

Figure 1: All three ROVs made by the RARL lab as at 2015. From left to right - Seahog, Robin and Challenger I

Before delving into the work surrounding the Seahog, an inexhaustive description of Unmanned Underwater Vehicles is provided for a general understanding of such a system and its variations.

1.1 Introduction to Unmanned Underwater Vehicles

The Unmanned Underwater Vehicle (UUV) has been around for over 150 years and has existed in various forms over the years. They have found uses in multiple industries: transportation, offshore oil field development and maintenance, seabed cable laying, sunken equipment or ordnance salvage missions, underwater sample collection, ocean life observation, sewer inspection etc. These vehicles have made the exploration of the ocean safer as they are able to reach depths that divers could not get to and removed the dangers associated with manned underwater vehicles. There are numerous variation of UUVs that exist in the market. Based on how they are being powered and their communication mode, it could either be tethered or untethered. When it is untethered, the vehicle is commonly referred to as an Autonomous Underwater Vehicle (AUV). The AUV is self-powered and operates without any input from a pilot by following a pre-programmed path or by determining the best path to fulfilling an objective. The tethered vehicle on the other hand is generally remotely controlled by a pilot(s) and thus referred to as a Remotely-Operated Vehicle (ROV). Power and communication are sent through a tether connected to a surface control unit. These vehicles (AUVs and ROVs) generally come with a combination of these sensors/components: video cameras, lighting unit, robotic manipulators, sonars, pressure sensors and localization and navigation units such as IMUs and Doppler Effect sensors. The bigger ROVs come equipped with a tether management system to ensure that the tether does not get in the way of the ROV while in operation. A combination of these features and its power output determines the class of a UUV. The vehicles that are made for observation with optional light sampling functionality are within the *observation class* while those that perform tasks such as cable laying and underwater site maintenance are within the *working class* [4]. The working class UUVs can still be sub divided into *heavy work* or *light work* depending on the power of their propulsion system, depth rating and the capability of their manipulators [4].

Figure 2: Left - ROV SIRIO [5], an observation class ROV; right - Triton XLS 150 [6], a working class ROV

As an example, Figure 2 and Table 2 highlight the properties of two different ROVs: the ROV SIRIO and the Triton XLS 150. This is just to demonstrate the kind of variation that exists within the commercial ROV sector.

Table 2: Comparison between two ROVs from different classes

	ROV SIRIO	Triton XLS 150
Size (L × W × H)	0.59 × 0.56 × 0.45 m	3.5 × 1.78 × 1.93 m
Dry weight	40 kg	4400 kg
Depth rating	300 m	3000 m

Offshore environmental activity regulators and government agencies have also found potential in UUVs. AUVs have been used as a tool for performing Environmental Effects Monitoring (EEM) specifically at oil fields in assisting with the Environmental Risk Assessment (ERA) [7]; especially as oil exploration and production continue to reach deeper and more remote waters [7]. EEM aims to study anthropogenic activities and help improve or add regulations to aid in minimizing the consequences of underwater activities on the marine ecosystem.

As alluded to at the beginning of this section, UUVs are still very relevant and even more so now to a host of disciplines fulfilling their activities offshore as new frontiers are being explored. However, due to the nature of the Seahog, the type of UUV that will be further discussed will be the ROVs.

1.2 Historical Background on ROVs

The earliest development of the ROV as it is known today is not very clear, although credit can be given to Luppis-Whitehead Automobile in Austria for making the PUV (Programmed Underwater Vehicle), a self-propelled torpedo which was completed and successfully demonstrated in 1866 [8] [9]. More relevant is Dimitri Rebikoff, who developed the first tethered ROV named POODLE in 1953 [8]. The subsequent development of the tethered ROV in the 1960s were funded by the United States Navy who were motivated by the need for robots capable of retrieving lost artillery and sunken submersibles from the seabed. This development culminated in the production of the US Navy's first series of ROVs called the Cable-Controlled Underwater Recovery Vehicle (CURV) [10]. Figure 3 shows the first in the series. Some of these ROVs were involved in some high profile salvage missions: CURV-I was used to retrieve a hydrogen bomb lost in the Mediterranean Sea in 1966, CURV-III was deployed in the rescue two pilots from a sunken manned submersible PISCES IV off the coast of Ireland in 1973 [10].

Figure 3: The US Navy's CURV ROV I, an 800 ft. (about 243 m) rated ROV is one of the earliest underwater ROVs [11]. This ROV was equipped with flux gate compass, depthometer, TV and still cameras, lights, altimeter, sonar and a variety of robotic grippers [10].

Further development later made were by the commercial sector, specifically in the area of offshore drilling. Working class ROVs were manufactured to help in the development and maintenance of oil fields [8]. The first commercial ROVs were made by Hydro Products in the US which were the RCV 225 and the RCV 150 (as seen in Figure 4). These ROVs were classed as observation or light work class ROVs as they were mainly used for oil field rig inspection and they (particularly RCV 150) were also capable of handling tools to perform maintenance work when necessary. With increasing venture into previously unreachable or unsafe sea depths, UUVs became a necessity in the 1980s for companies dealing in the oil field management [8]. Ever since then, various companies have become involved in the development of commercially available ROVs, supplying them to the marketplace at different price ranges depending on the size and capabilities of the vehicles.

Figure 4: "eye ball" ROV RCV 225 (left) and RCV 150 (right) by Hydro Products [8]

The cost of ROVs varies depending on the system specification: for a simple portable tethered ROV made for personal use and rated for low depths (less than 80 m), prices starting from about $1,000 are typical while a medium-sized ROV with basic functionality, without manipulators and with functional ratings similar to the Seahog costs around $65,000 [12] (prices stated as at April 2018).

1.3 The Seahog's Status

As stated in earlier in this chapter, the design of the Seahog is a modular open-frame design. The following figure shows the breakdown of each subsystem of the Seahog.

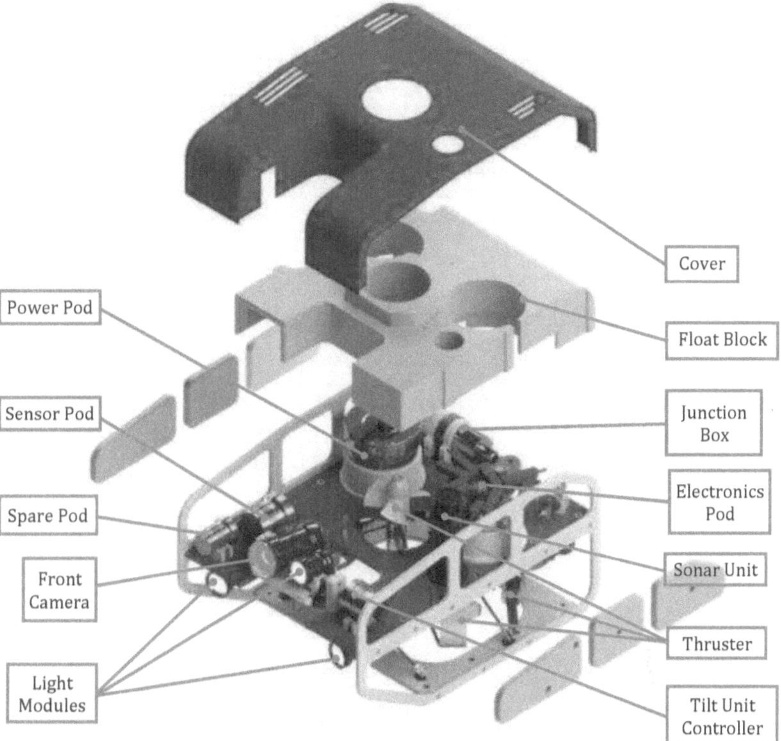

Figure 5: The Seahog and all its modules and features

The following table explains the functions and operational status of each module as at the commencement of this project.

Table 3: Seahog's modules, their respective functions and working status

Module	Function	Working status
Float block	Adds buoyancy to the system	Manufactured and installed
Cover	Adds extra level of protection for some of the modules. Adds to the Seahog's aesthetics	Yet to be manufactured
Subsea junction box	Connects surface power and communication to Seahog through tether's copper and fibre optics cables	Operational
Power pod	Converts surface supply of 400 V DC into 5 V, 12 V, 15 V and 48 V lines	Operational
Electronics pod	Distributes power and serial communication to each module	Operational
Thruster modules	Provides vertical and horizontal thrusts to the Seahog	All operational
Front and rear facing cameras	Acts as a navigation tool by providing live feed video footage to the surface control	All operational
Light modules	Provides illumination	All operational
Sonar unit	Helps with navigation by providing information about external objects' proximity to Seahog	Operational

Module	Function	Working status
Sensor pod	Contains the pressure sensor and IMU to be used for depth and attitude sensing respectively	Uninstalled and untested
Spare pod	Pod for extra utilities	Unused
Tilt unit controller	Controls the front camera's tilt position	Uninstalled and untested

1.4 Project Objectives

The focus of this project was to get the Seahog to a functional state where feedback and subsequently some control is achievable. The objectives of this project were therefore to procure or design, test and install the remaining pertinent modules yet to be installed namely the *sensor pod* and *tilt unit controller*. The integration of these subsystems with the overall Seahog setup with regards to power and communication is also of significance if the Seahog is to reach a state of full functionality and controllability. The current Seahog control model on Simulink was based on some hydrodynamic coefficients, specifically the added mass, that had not been verified, especially because of the simplifications made in the approach taken in estimating the [2]. This was therefore another priority of this project: to investigate the Seahog's hydrodynamic model.

The objectives can thus be outlined as:

1. Design and test an attitude estimating subsystem to be implemented on the Seahog.
2. Design, test and integrate the tilt unit controller with the front camera and light system.
3. Investigate the hydrodynamic added mass model of the Seahog used in the simulation model.

This book is broken into three parts: part 1 is the introduction (this current chapter), Part 2 deals with the attitude work and part 3 deals with the hydrodynamic added mass work. The next three chapters present comprehensive detail of work done in fulfilling objective 1. This is followed by another two chapters of literature review and work done in improving the Seahog's hydrodynamic mass model, which is objective 3. Objective 1 was viewed as the most neglected and pertinent part of getting the Seahog to a reasonable functional state which is the reason for it requiring the most time dedication and effort. The course of each segment will be to provide some background information at the beginning in the form of an introduction section followed by the groundwork of the work to be done which is then all concluded with an analysis of the results from the work performed. Work done in fulfilling objective 2 can be found in the appendix section as the groundwork required for setting up and successfully implementing a tilt unit controller had been previously established and documented by Hope [13]. The task here was therefore heavily implementation-based about rather than design-based.

2 LOCALIZATION OF THE SEAHOG

2.1 Introduction

This chapter is a precursor to fulfilling objective 1: *to design and implement a functional attitude estimation system for the Seahog.* The concepts of *localization* and *attitude* (in the following chapters) are subsequently explained with reference to the Seahog.

This chapter is divided up into sections as follows:

- The reference frame of the Seahog is described in detail as it is relevant to this and subsequent chapters.

- A brief discussion on localization and the technologies used for the localization of underwater vehicles.

2.2 The Kinematics of the Seahog

As the Seahog is to be controlled aboard a surface vessel that is assumed to be ideally stationary, its localization needs to be defined relative to a known position or reference frame. This section of the book explains the process and motivation for the defined sets of reference frames subsequently employed.

The Seahog is a six degree-of-freedom (DOF) vehicle: three linear and three rotational. Figure 6 shows the notations used in representing each of the Seahog's DOF.

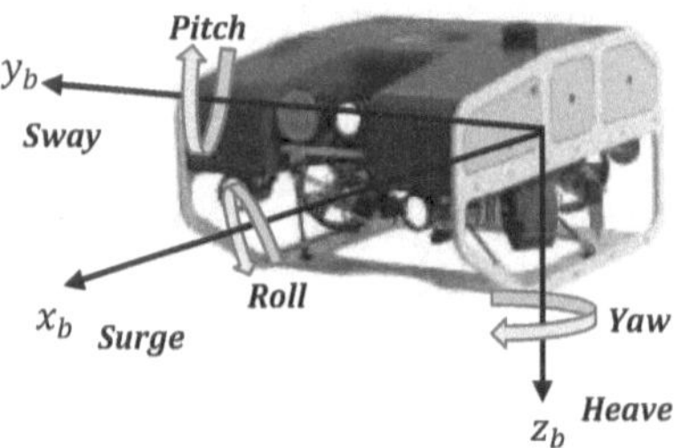

Figure 6: Six DOF motion definition with the body-fixed frame

The reference frame of axes x_b, y_b, z_b represent a *body-fixed* coordinate system fixed to the moving Seahog. This is the frame from which the Seahog's linear and angular velocities can be described.

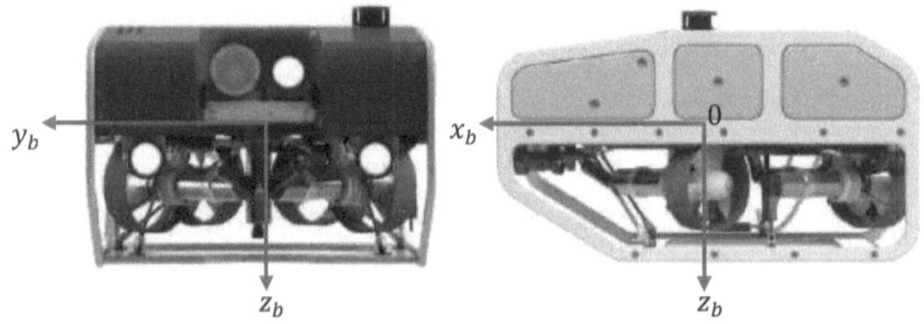

Figure 7: Body axes system of the Seahog

With regards to the Seahog's body-fixed reference frame, the point of origin was chosen to be the intersection of the vertical thruster's axis and the plane of the bottom surface of the base frame (as seen in Figure 7) [2]. The reason for choosing this point of intersection of the aforementioned axis and plane rather than the Seahog's centre of mass was to create a partial three-plane of symmetry frame system, which simplifies the Seahog's geometry for subsequent dynamic analyses in this and the next chapter. Due to the low speed nature of the Seahog's motion, the acceleration of a point on the earth can be neglected for marine vehicles since the earth's motion hardly affects slow moving marine vehicles [14], therefore, the coordinate system attached to the surface vehicle can be taken as the inertial *earth-fixed* reference frame. This is displayed in Figure 8.

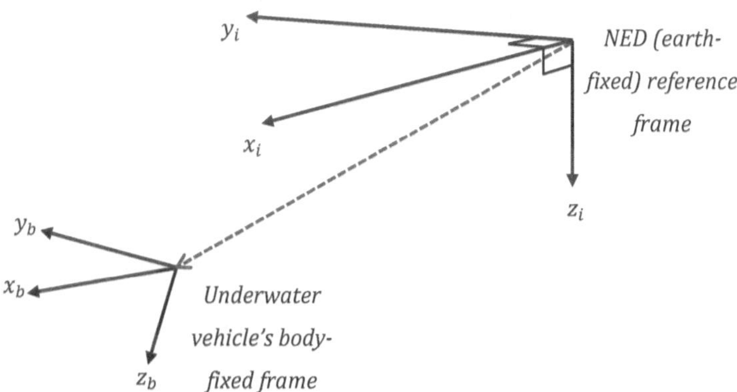

Figure 8: The reference frames of the underwater vehicle and surface vehicle with respect to each other

The earth-fixed frame can also be referred to as the *North-East-Down (NED)* frame. In this reference frame, on the relevant point of the earth (the position of the surface vehicle here), the x_i-*axis* points towards true north, the y_i-*axis* points towards the east while z_i-*axis* points downwards perpendicular to the tangent plane at that point on the earth. The Seahog's position and attitude can be represented relative to the NED frame. The NED frame is also classified as a *local geographic* frame since the NED frame is allowed to be applied on a flat ground due to the relatively short travel distance between the surface vehicle and the underwater vehicle [15]. This coordinate system is also employed in the navigation of aircrafts and can be deployed as an *ENU (East North Up)* frame if required [15].

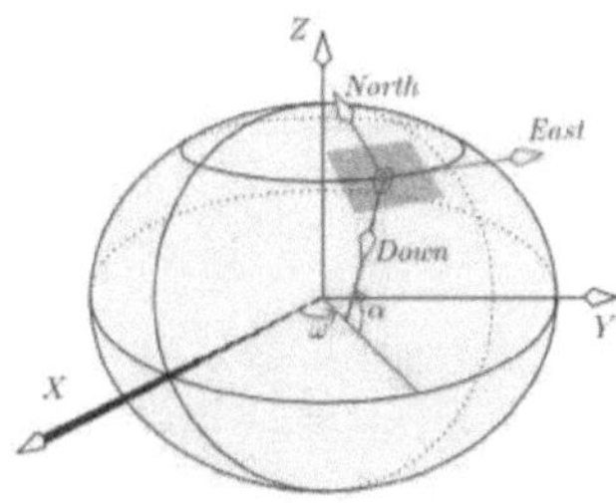

Figure 9: The NED frame is on a plane tangent to the earth's surface. The XYZ coordinate system is the earth centred, earth fixed (ECEF) frame which

**rotates with the earth relative to a non-rotating earth centred inertial (ECI)
frame, but as explained earlier, is being taken as inertial in its application
with respect to marine vehicles [15]**

The following table displays the notation used in this documentation to
represent the velocity, position and force/moment in each degree of freedom.
The symbols used here is in keeping with the standard designations defined by
the Society of Naval Architects and Marine Engineers (SNAME) [16].

Table 4: The Seahog's six DOF motion and dynamics notations

DOF	Linear and angular velocities	Forces and moments	Linear and angular positions (relative to the inertial frame)
x_b-axis	u	X	x
y_b-axis	v	Y	y
z_b-axis	w	Z	z
Rotation about x_b-axis	p	K	ϕ
Rotation about y_b-axis	q	M	θ
Rotation about z_b-axis	r	N	ψ

The following vectors can now be used to describe the Seahog's motion using the
notations provided in Table 4:

$$\eta = \begin{bmatrix} \eta_1 \\ \eta_2 \end{bmatrix}; \qquad \eta_1 = \begin{bmatrix} x \\ y \\ z \end{bmatrix}; \qquad \eta_2 = \begin{bmatrix} \phi \\ \theta \\ \psi \end{bmatrix}$$

$$v = \begin{bmatrix} v_1 \\ v_2 \end{bmatrix}; \qquad v_1 = \begin{bmatrix} u \\ v \\ w \end{bmatrix}; \qquad v_2 = \begin{bmatrix} p \\ q \\ r \end{bmatrix}$$

$$\tau = \begin{bmatrix} \tau_1 \\ \tau_2 \end{bmatrix}; \qquad \tau_1 = \begin{bmatrix} X \\ Y \\ Z \end{bmatrix}; \qquad \tau_2 = \begin{bmatrix} K \\ M \\ N \end{bmatrix}$$

η is the linear and angular position vector with coordinates in the earth-fixed frame (NED), $\boldsymbol{v}$ is the linear and angular velocity vector with coordinates in the body-fixed frame and τ is the force and moment vector with coordinates in the body-fixed frame [14].

2.3 Underwater Localization

To pilot an ROV underwater, there are quantities that are paramount and need to be monitored for the sake of accurate navigation. These quantities are employed in designing simulation and control algorithms for underwater vehicles. This section was deemed relevant to estimating the attitude of an underwater vehicle as attitude falls within the scope of localization.

Localization can be defined as the process of defining the position and attitude of a vehicle relative to a predefined or inertial position. Similar navigation and localization methods are applied to both underwater ROVs and AUVs. It is important to have localization data for the Seahog as this would be used in fine-tuning the Seahog's kinematic and dynamic simulations and the subsequent design of its controllers.

2.3.1 Underwater Localization Technologies

The tractable parameters relevant to underwater localization are the velocity, depth, position and attitude. Some of these quantities can be directly measured while others can only be estimated using sensor data in an algorithm.

2.3.1.1 Velocity

The *Doppler Velocity Log (DVL)* is a commonly used sensor which estimates linear speed. This is achieved by a transducer on the vehicle outputting an acoustic wave at a specific frequency and measuring the frequency shift of the response signal received from a responsive surface which can be the seabed or the surface. It is capable of measuring linear velocity in the surge, sway and heave directions by using at least three acoustic beams pointing in different directions which are then mapped onto an XYZ Cartesian frame [17]. It is also more accurate when tracking velocities relative to the seabed (this is referred to as bottom-tracking) rather than the surface because of potential currents which would negatively affect such measurements [17]. DVLs are typically large in size (one of the smallest commercial DVLs being a 300 m rated DVL of height 158 mm × diameter 114 mm [17]) and relatively expensive [18]. There is a trade-off between frequency and the range a DVL can travel from a responsive surface: the higher the acoustic signal frequency, the shorter the range [17]. Another velocity sensing device is the *Electromagnetic log (EM log)*. Where the DVL measures the vehicle's speed relative to an inertial position, the EM log measures the speed of water relative to the moving vessel. It operates by generating current in a coil proportional to the flowrate of water measured by a flow sensor [19]. A big disadvantage of the EM log is that speed is relative to water rather than an inertial point or ground. Boundary layer also plays a huge part in the measured speed as water on or close to a vessel's surface is slowed down due to surface friction [19].

2.3.1.2 Depth

Estimating the depth travelled by an underwater vehicle is generally achieved using a *pressure sensor*. The accuracy and output rate of the depth estimate is dependent on the data output rate, precision and accuracy of the pressure sensor installed. Acoustic transducers could also be used to estimate depth.

2.3.1.3 Position

The position of an underwater vehicle can be estimated or measured using a combination of several transponders, sensors and methods. The most effective positioning system for an underwater vehicle would depend on factors such as vehicle's operational environment, on-board sensors, vehicle type - autonomous or remotely controlled and size of the vehicle.

- **Dead reckoning**: This is a relative positioning technique that involves the tracking of a vehicle's position relative to some inertial reference frame. This approach involves the use of known sampling period, IMU sensors and a velocity log to estimate current position relative to a known starting point. The accuracy of this approach is hugely dependent on the precision and accuracy of the sensors and the algorithm used in obtaining the position estimate. There is however a high possibility of estimate drift due to the accumulation of sensor errors overtime. Typical sensors used in dead reckoning are the IMU (for attitude estimation) and the DVL.

- **Global Positioning System (GPS)**: The GPS is a device capable of receiving *radio* signals from at least three transmitting satellites and estimating its location through a process of *trilateration*. Radio signals and *electromagnetic* signals in general are however quite weakly propagated through seawater because of its high electrical conductivity. The higher the radio signal's frequency, the greater its attenuation in seawater [20]. The signal would require a lot of amplification for it to be useful for underwater localization. This renders the GPS rarely useful in deep water navigation. Electromagnetic signals have however been found useful in some areas of underwater navigation; in the areas of short range applications and wireless communication between UUVs to surface vessels and UUV to UUV in fresh water. This is because fresh water has electrical conductivity typically far less than that of seawater, thereby causing less signal attenuation [20]. Low frequency radio waves can also be used on shallow water vehicles . There are some key advantages of electromagnetic-based sensors over acoustic-based sensors in underwater navigation [21]:

  o No known effects on marine life unlike the acoustic transducers which have been found to affect marine mammals.

  o High bandwidth and data rates.

  o Compact unit.

  As at 2018, the Defence Advanced Research Projects Agency (DARPA) in conjunction with BAE Systems, Raytheon BBN and Draper Laboratory have been working on a project called *Positioning System for Deep Ocean Navigation (POSYDON)* which aims to develop a new navigation system for underwater drones using a combination of acoustic and GPS signals [22] [23]. If this succeeds, it will have the potential for deployment in general purpose deep water UUVs.

- **Acoustic positioning systems**: This approach involves the use of acoustic signals in the positioning of an underwater vehicle relative to a known reference point. The popular methods being considered here are the *Ultrashort Baseline (USBL), Short Baseline (SBL)* and *Long Baseline (LBL)*. The baseline approach is essentially a positioning technique in which the UUV determines its distance relative to three or more *beacons* or *transceivers* by measuring the phase drift of the returning acoustic signals sent out by a transducer. SBL and USBL are typically deployed such that the vehicle localizes itself relative to a surface vessel or seabed. In the USBL approach, also called the *Super Short Baseline (SSBL)*, multiple transponders (beacons) are positioned closely together along the hull of a surface vessel or on the seabed. The transponders are positioned over a small area of the seabed or along the ends of the hull of a ship in the case of the SBL [18]. Although the seabed can be used as a reference point for the UUV's transducer, the beacons are typically deployed along the surface vessel's hull. LBL on the other hand is such that the beacons are placed over a wide area of the seabed. The vehicle then uses triangulation to localize itself relative to these beacons. Out of the three acoustic positioning systems mentioned, a high frequency LBL produces the most accurate position estimate which can be as fine as a few centimeters [17]. Though the USBL is the simplest to implement, it is also the most susceptible to noise.

Acoustic-based signals are therefore preferable in underwater navigation and localization over electromagnetic-based transducers. However, there are some draw backs to acoustic systems that need to be considered when designing or selecting a positioning technique for a UUV. Some of these are [18]:

  o Small bandwidth.

  o Sound travels at about 1500 m/s in seawater (at room temperature) which is much slower than the speed of an electromagnetic wave which travels at the speed of light.

o Low data rate.

o Susceptible to multi-path when the vehicle is in a short range of the sea's surface or seabed due to reflections from the surface or seabed respectively [20].

There are other acoustic signal-based systems such as the *sonar*; which can be implemented in two forms: imaging and ranging, and the *GPS intelligent buoys (GIBs)* which is similar to the LBL but with GPS-tracked beacons installed at the surface of the sea [18].

- **Simultaneous Localization and Mapping (SLAM)**: This computational philosophy like the name suggests is the approach in which the pose (position) of an underwater vehicle is determined simultaneously as environmental mapping is performed. This dependency between localization and mapping exists as a vehicle needs an accurate estimate of its pose in order to be able to map its environment while simultaneously requiring a consistent environment map to determine its position [24]. SLAM uses an algorithm such as a Kalman filter and real world observation from sensors such as to speed sensors, cameras, optical sensors, range finders etc to track a vehicle's position within an environment. This philosophy can be implemented in both AUV types (as *active* SLAM, where the vehicle uses information gathered to plan navigation trajectory autonomously) and ROVs (as *passive* SLAM, where environmental information is used by the pilot to navigate the vehicle) [25]. There are different approaches to implementing SLAM. Some of the common SLAM techniques and their counterparts are:

  o *Feature-based SLAM vs View-based SLAM*: A feature-based SLAM is the type in which the pose of the vehicle, relative to specific salient features are maintained in state-space [18]. This approach leads to a sparse mapping of the vehicle's environment and is therefore computationally low cost, more efficient and fast when compared to the *volumetric* or view-based SLAM. View-based SLAM on the other hand produces a dense mapping of an environment without ignoring any environmental data which may lead to superior results in certain situations [25]. The view-based SLAM is more suited to environments that are unknown or under-defined.

- o *Full SLAM vs Online SLAM*: Online SLAM recovers the current pose estimate of the vehicle while full SLAM estimates the entire path of the vehicle together with the map [25] [18].

There are other contrasting SLAM dimensions such as *static* vs *dynamic*, *topological* vs *metric* [25]. Most SLAM implementations are based on three main paradigms or algorithms: Kalman filter, particle filter and graph-based [25]. A major pitfall of SLAM is that most of its techniques deal with static environments when in reality, underwater vehicles operate in dynamic environments [25]. Some important performance factors to consider when implementing SLAM for an underwater vehicle are:

- o Computational memory and speed costs

- o Operational requirements

- o Sensors available

The descriptions provided in this subsection have been a broad and inexhaustive look at the navigation philosophy of *SLAM*. For a more comprehensive study of the SLAM techniques, read the chapter on SLAM by Thrun and Leonard [25].

2.4 Localization Systems Employed in Modern ROVs

The following are some examples of localization technologies currently deployed in modern ROVs by some of the UUVs industry's leading manufacturers. The following ROVs were produced by SAAB AB [26], one of the leading UUV makers in the world.

- Sea Wasp: Unveiled in 2015, this ROV is used by defence forces for Improvised Explosive Device (IED) removal. It employs the fusion of the IMU, DVL, compass and a multi-beam sonar for localization.
- Double Eagle MK II/MK III: These are ROV/AUV hybrids capable of being deployed untethered. They are used by navies for mine reconnaissance and disposal. Their current models employ USBL, MEMS sensors, speed log and

DVL. Inertial Navigation System (INS) and GPS are optional sensors that can be added based on function specifications.

- SUBROV: An ROV made for submarine deployment employs DVL and MEMS sensors for localization with INS, GPS and speed log as optional additional units.

The observation/inspection class ROV XLe Spirit, introduced in 2018 by Forum Energy Technologies primarily uses an AHRS unit for attitude (relative angular position) sensing [27]. There are optional units that could be installed such as the north-seeking gyroscope, multi-beam sonar and magnetic compass depending on functional specifications. Omni Maxx is a 3000 m depth rated observation class ROV by Oceaneering International [28]. It employs the VN-100 Rugged, a high performance IMU unit which includes a 3-axis, gyroscope, accelerometer and magnetometer for heading sensing.

3 An Introduction to Attitude

The attitude or orientation of a vehicle is its angular position relative to an inertial reference frame. Attitude encompasses the roll, pitch and yaw angles. The attitude measuring device is therefore one that measures the tilt (roll and pitch) and yaw angles relative to a reference frame of a UUV. Commercial units such as the tilt sensor or inclinometer can be used for tilt while a gyrocompass can be used for heading measurements. Note that while yaw and heading are similar and refer to the third attitude angle, yaw is stated relative to an inertial frame while heading is typically stated relative to the earth's frame, i.e. stated with respect to the earth's magnetic or true north. In this document however, yaw is the term in use as attitude is taken to be solely relative to a specified inertial frame.

Measurements from certain sensors can be used to estimate a vehicle's attitude. There are units comprising of a number of these sensors. Such a unit is generally composed of an accelerometer, gyroscope and magnetometer. This unit is commonly referred to as an *IMU (Inertial Measurement Unit)*.

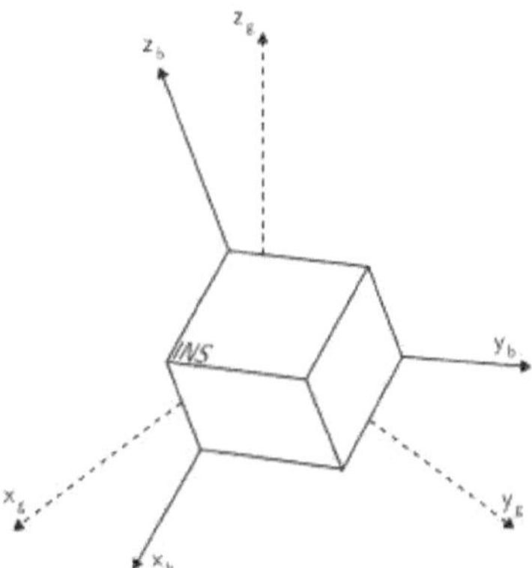

Figure 10: The x_g, y_g and z_g axes is the global frame representing the reference frame, while axes x_b, y_b and z_b is the body-fixed frame [29]

The most common IMUs are six DOF units usually consisting of three-axis gyroscopes and three-axis accelerometers. The accuracy and reliability of the IMU unit would depend on the type and characteristics of each sensor. IMUs can be applied in two different configurations namely the *stable platform system* or the *strap-down system* [29]. The stable platform setup is such that the IMU, set within a gimbal system, is placed on a platform. The aim of this setup is to keep the platform's frame aligned to a global frame, i.e. to stabilize the platform [29]. The motion of the gimbals (which could be controlled by motors) act to counteract any external rotation in order to keep the platform stable. This setup is mainly employed in systems requiring stabilization such as cameras mounted on drones or helicopters. On the other hand, the IMU could be mounted on a device in the strap-down setup. Any angular rate or linear acceleration is captured by the embedded gyroscope or accelerometer respectively. This setup is employed when the attitude or position information of a device or system is required. It is employed in motion capture and essentially any system requiring attitude tracking.

The output signals from IMUs can be digital - this case requires data to be read through a specific communication protocol or analogue - voltage/current is read and converted to angular rate based on the IMU's parameters. The average IMU is capable of sensing six degrees of freedom through three-axis gyroscope and three-axis accelerometer on-board. An IMU can have an additional three-axis magnetometer installed. This sensor set can be used to implement a sensor fusion system referred to as the *Magnetic Angular Rate and Gravity (MARG)* sensors. The MARG sensor system additionally employs the earth's magnetic field to estimate attitude. The following is a brief look at the main IMU sensors.

- Gyroscope: The gyroscope is a device that measures the angular rate or velocity of a vehicle about an axis relative to an inertial frame. A three-axis gyroscope installed on a UUV measures the angular rate about each body-fixed axis. A gyroscope comes in two general types namely mechanical vibration-based and optical-based. The mechanical vibration-based gyroscope operates with an oscillating mass suspended within a spring system [18]. The coriolis force on the mass resulting from the gyroscope's rotation is then used to calculate the angular rate. This type of gyroscope is commonly deployed as a *Micro-Electro Mechanical Systems (MEMS)* gyroscope in UUVs. MEMS inertial sensors are commonly used in a lot of systems today such as the digital cameras, smartphones and virtual reality sets. This is mainly due to the advantages they offer, some of which are their small size, low production cost, low power consumption, robustness and short start time [30]. The optical-based gyroscope has two modes of operation. The *Ring Laser Gyroscope (RLG)* uses mirrors or prisms at the edges to direct a light beam generated by a laser along straight-line light paths [31]. The *Fibre Optic Gyroscope (FOG)* also uses a laser as its source of light, but this light beam is passed through glass-fibre loops [31]. In both optical methods, two light beams are sent out in opposite directions from a similar starting point. FOG measures the angular rate from the phase difference between the two beams caused by a rotation of the gyroscope. This phase difference which is due to the difference from distances travelled by each light beam to reach the starting position of the rotated path is referred to as the *Sagnac effect* [31]. RLG measures angular rate from the fringes of the interfering clockwise and counter-clockwise beams [31]. All gyroscopes are susceptible to drift which is a result of measurement error accummulation over time [31]. Optical gyroscopes are less sensitive to temperature changes. They also provide more accurate measurements and the least drift over time, with drift as small as 0.0001 °/hour while MEMS gyroscope can have drift as small as 0.01 °/hour [30]. MEMS gyroscopes are however cheaper, smaller and consume less power than optical-based gyroscopes [18]. Some factors that affect a gyroscope's performance are temperature sensitivity, bias, drift, precision and angle random walk or noise.

- Accelerometer: This is a device that measures specific force which includes effects from external and gravitational forces acting on a vehicle. This is achieved in two general ways. One method is to implement the accelerometer *mechanically* by employing a mass-spring system similar to the MEMS gyroscope's vibration-based operation. Force is measured by the amount a spring is displaced by the mass in response to an external force [31]. Another approach is the use of the *piezoelectric effect*. This effect is an electrical property of some materials such as silicon to induce a detectable electrical charge when stretched or compressed. Piezoelectric accelerometers are primarily composed of a *piezoelectric crystal* (usually quartz crystals), a *proof mass* which acts to induce stress in the crystal in response to an external force and a conditioning IC. Benefits of a piezoelectric accelerometer over a mechanical accelerometer are its lack of moving parts, low output noise and self generation of power [32]. Accelerometers can be used to estimate tilt angles when the vehicle is in steady state, i.e. the gravity is only force on the body. Accelerometers are quite sensitive to vibrations which are a source of high frequency noise on measurements during underwater navigation. This makes their measurements unreliable in the short term when integrating their measurements when estimating pose.

- Magnetometer: This device measures the total local magnetic field strength. A magnetometer belongs to one of two main types: *vectored* or *scalar*. The scalar magnetometer produces the absolute magnetic field strength value while the vector magnetometer consists of three or more magnetic field sensors capable of producing the magnetic field measurement in a vector components form [33]. Several magnetic sensors can be used in developing a vector magnetometer such as hall sensors, fluxgate sensors and magneto-resistive magnetometers [33]. Some scalar magnetometer types commonly used are the proton precession, Overhauser effect and optically pumped magnetometers [33]. Some disadvantages of the scalar magnetometers are their bulky size, high power consumption and low magnetic field resolution [33]. MEMS magnetometers are commonly found in MARG IMUs. They are typically vector magnetometers that work on the principle of Lorentz force effects detection: a change in voltage caused by a magnetic field is measured. They are primarily used for detecting earth's magnetic field when deployed in UUVs. Magnetometers are especially useful and preferable for estimating the UUV's heading position relative to the earth's magnetic north. The major shortcoming of the magnetometer is its high sensitivity to external influences in magnetically dirty environments, i.e. where magnetic fields other than the earth's are prevalent.

3.1 The Seahog's IMU System

The STEVAL-MKI062V2 (iNEMO) is a demonstration board made by STMicroelectronics. Amongst a few other sensors such as a PCB temperature sensor, the iNEMO possesses a nine DOF MEMS IMU which was pre-selected to be the Seahog's IMU by the team that previously worked on the Seahog. The iNEMO was deemed sufficient in fulfilling the objective of this project. The iNEMO is about 40 mm × 40 mm × 2 mm in size. An external pressure sensor for depth measurement was also preselected. Measurements from the pressure sensor were fed to the iNEMO's microprocessor (MCU) through an IO (Input/Output) pin because the pressure sensor outputs a continuous analogue signal. Table 5 contains the properties of the pressure sensor and iNEMO's factory calibrated IMU sensors.

Table 5: IMU's sensors parameters [34]

Type (Sensor)	Range	DOF	Operating voltage range	Noise density	Operating temperature range	Output data rate (ODR) range	Interface
Accelerometer (LSM303DLH)	±2 g, ±4 g, ±8 g	3	2.5 – 3.3 V	$218\,\mu g/\sqrt{Hz}$	-30 – +85 °C	50 – 1000 Hz	I²C
Magnetometer (LSM303DLH)	±1.3 – ±8.1 gauss	3	2.5 – 3.3 V	5 mgauss (RMS)	-30 – +85 °C	0.75 – 75 Hz	I²C
Roll and pitch rate Gyroscope (LPR430AL)	±300 dps	2	2.7 – 3.6 V	$0.018\ dps/\sqrt{Hz}$	-40 – +85 °C	140 Hz	Analog output
Yaw rate gyroscope (LY330ALH)	±300 dps	1	2.7 – 3.6 V	$0.014\ dps/\sqrt{Hz}$	-40 – +85 °C	140 Hz	Analog output
Pressure (SICK PBT - 6042443)	400 m	1	14 – 30 V		0 – +80 °C		Analog output

The following values are the empirical mean bias values of the accelerometer and the gyroscope.

Table 6: Sensor bias

	x-axis	y-axis	z-axis
Accelerometer	-3.024×10^{-3} g	-3.5871×10^{-3} g	-47.3899×10^{-3} g
Gyroscope	1.6190 °/s	3.4399 °/s	-0.0042 °/s

Measurements from these sensors were read and processed on the MCU on-board the iNEMO. The following table details its important features.

Table 7: iNEMO's microprocessor

STM32F103RET microprocessor	
Max clock frequency	72 MHz
Flash memory	512 KB
SRAM memory	64 KB
Voltage supply	2 – 3.6 V
Input/output (IO) pin voltage range	0 – 3.6 V
General Purpose IO (GPIO) pins	51
Timers	8
Analogue-to-digital converters (ADC)	3 (12-bit max. resolution)
Communication protocols	I^2C, SPI, USART, USB, SDIO

3.2 iNEMO's Coordinate System

The following figure shows the measurement output coordinate system that applies to each IMU sensor due to their position on the PCB and the iNEMO's reference frame. When performing sensor readings and calibrations, the gyroscope's mismatching y-axis was matched to that of the IMU's to ensure accurate attitude computation.

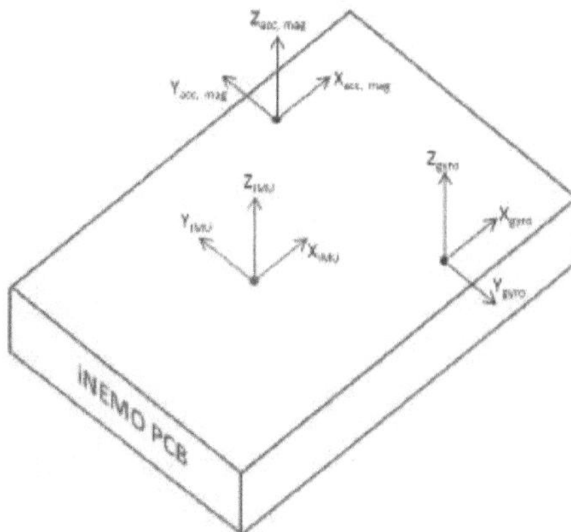

Figure 11: The reference system of the gyroscope (gyro), accelerometer (acc), magnetometer (mag) and the overall IMU reference frame

The IMU and Seahog's reference systems were not aligned in all axes as shown in Figure 12. The positive y and z axes were in opposite directions. The signs of the pitch and yaw results from the attitude's computation were therefore flipped to ensure uniformity with the Seahog's body-fixed frame.

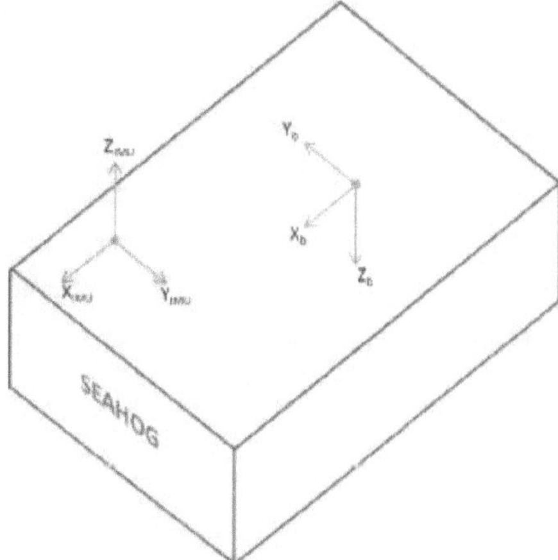

Figure 12: Axes X_b, Y_b and Z_b are the Seahog's NED body reference frame while X_{IMU}, Y_{IMU} and Z_{IMU} axes frame represents the IMU's reference frame

3.3 Exclusion of the Magnetometer

Due to the proximity of the IMU to magnetic couplings in the vertical thruster, the two front thrusters and the magnetic coupling in the tilt unit, the data from the magnetometer was severely affected. There are algorithms for compensating for distortion to the magnetometer's magnetic field measurements. Most of the methods put less significance on the magnetometer's measurements after a defined magnetic field threshold has been crossed. Some examples of this being implemented can be found in works by Fan et al. [35] and Yadav and Bleakley [36]. A threshold could be a combination of the measured earth's magnetic inclination and field intensity. It was however expected that the local magnetic field around the IMU would change as the motion of the thrusters and tilt unit varied. There were also current carrying cables on board which generated magnetic fields of their own. There is also the possibility of magnetic field superimposition from the environment in which the Seahog would operate.

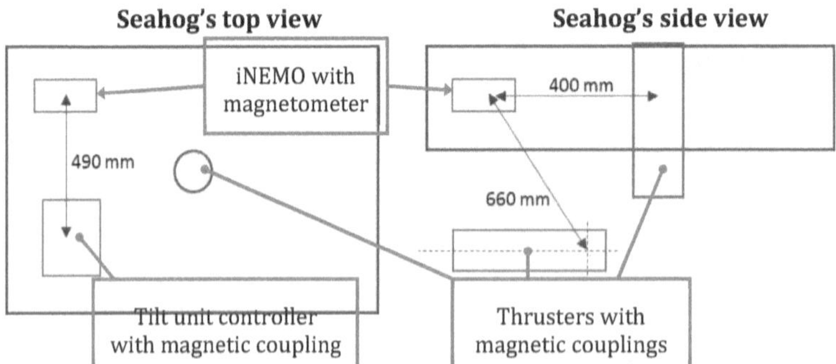

Figure 13: The distances between the magnetometer and the vertical and front thrusters

This ultimately led to the exclusion of the magnetometer from the IMU's sensor outputs to be used in estimating the attitude.

3.4 Attitude and Heading Reference System

The Attitude and Heading Reference System (AHRS) is a collection of sensors and systems that provides attitude (roll, pitch and yaw) and heading information of a vehicle in 3D space. Where yaw states angular displacement from an inertial frame, heading is typically based relative to a compass direction, i.e. true north. The accuracy of the attitude provided by the AHRS is dependent on the precision of the sensors involved, the computational capacity of the embedded processor and the accuracy of the sensor fusion algorithm employed.

The attitude of the Seahog can be provided at a snapshot by each of the main IMU sensors; the issue with direct gyroscope rate output integration is that it drifts overtime, while the accelerometer's outputs are noisy, cannot be used to estimate yaw and are only useful in steady state. The shortcomings of the various individual sensors are the reason for *sensor fusion*. A sensor fusion algorithm combines data from all sensors and ensures an optimal attitude estimate output.

There are several ways in which attitude can be represented both computationally and communicatively. The following subsection looks at the various attitude representations.

3.4.1 Attitude Representation

There are different methods of representing rotation in 3D space. Some of the factors to consider when choosing a representation are their computational costs, numerical stability and user interaction [37]. The following is a description of some common vehicle attitude representations.

Rotation Matrix

This involves the use of a transformation matrix to represent any arbitrary rotation about a reference frame in 3D space. The rotation matrix can represent multiple rotations. For example, take a rotation about a body-fixed x-axis to be R_x, about the y-axis to be R_y and about the z-axis to be R_z. A rotation sequence XYZ is therefore the matrix product $R = R_z R_y R_x$. This implies that we rotate about the x-axis first, then the y-axis and finally the z-axis. Note that the order of multiplication matters as the matrix R is not commutative.

Fixed Angles

This is an extrinsic representation of attitude. Extrinsic rotations involve a sequence of rotations that are relative to a non-moving reference frame while intrinsic rotations are a sequence of rotations about the moving body-fixed frame. For example, an intrinsic rotation sequence *XYZ* (*xy'z''*) is represented by the matrix product $R_z(\theta)R_y(\beta)R_x(\alpha)$. Firstly, the x-axis is rotated through an angle α, then the new y-axis (*y'*) is rotated by β and finally the new z-axis (*z''*) is rotated by θ. A drawback to using fixed angle representation is the occurrence of *gimbal lock*, i.e. the loss of a DOF when two axes coincide after a rotation of 90 ° or its integer multiple about a third axis [37].

Euler Angles

Euler angles are three numbers representing any arbitrary rotation in 3D space. Where the *Tait-Bryan* angle representation involves rotations about three distinct axes, the *proper Euler* representation involves rotations about two distinct axes. Each representation has six possible combinations for a 3D rotation. The following are the six possible rotation sequences for each configuration using extrinsic notations:

Proper Euler angle sequences: xyx, xzx, yxy, yzy, zyz, zxz

Tait-Bryan angle sequences: xyz, xzy, yxz, yzx, zyx, zxy

These sequences are therefore sufficient in representing any rotation in 3D space. These sequences are usually collectively referred to as Euler angles, which is the reason for the repeating sequences being referred to as the proper Euler. The intrinsic version of the Tait-Bryan sequence *zyx* is commonly used in aerial and marine navigation. The Euler angles for this sequence are called yaw, pitch, and roll respectively for marine navigation. Intrinsic Euler angle rotation sequences have a reverse relationship with fixed angle representation. For example, *xyz* (Euler) = *zyx* (fixed) . This means that an Euler rotation sequence can be represented extrinsically using fixed angles. A major drawback to using Euler representation is the possible occurrence of gimbal lock. They are also computationally tasking.

Quaternions

Quaternions are an extension of complex numbers onto 4D space. A rotation is represented by a four-element vector. Attitude is essentially represented by the shortest rotation from the reference frame to the body-fixed frame of the unit quaternion vector lying on the surface of a hypersphere. It has the advantage over fixed and Euler representations of having the full rotation range about all axes without the risk of gimbal lock. The following vector equation is of a quaternion q and its components. This is also the standard form adopted throughout this report. It is made up of a scalar part q_o and three vector parts.

$$q = q_0 + iq_1 + jq_2 + kq_3$$

Quaternions are harder to intuitively understand. They are simpler to implement computationally especially when normalization is required as it is easier to normalize four numbers than a nine-element matrix [15]. There are also no singularities with quaternion computations.

3.4.2 Methods of Attitude Estimation

The world of attitude estimation is an extensive one. There are multiple approaches to obtaining a vehicle's attitude. The most commonly used methods were found to be variations of *Bayes' filters*, particularly, the *Kalman filter*. Bayes' filtering is a recursive method generally involving a cycle of *prediction* and *update* stages for estimating the *state* of a dynamic system using observations or measurements from said system [38]. Due to time constraints set for tasks in this project, attention was given to only two implementations of Bayes' filters and a non-Bayesian filter. The characteristics of these filters will be described and compared. The Bayes' filters will mostly be discussed from a *state-space* view point rather than a statistical one as would be evident from the notations used in the ensuing discussions.

3.4.2.1 Kalman Filter

The Kalman filter (KF) is an optimal *linear* state estimator for dynamic stochastic systems. It optimises the state estimate by *minimising* the *square difference* between the *predicted state* and the *actual state*. It estimates the current state of the system by predicting the state a priori using a predefined system model. It can be setup to provide the most optimal state estimation by using real life system measurements to better calibrate the estimate from the system model. The Kalman filter algorithm can be represented using state-space models. The following discrete state-space equation represents a typical linear system, whose state x_k at a time step k is a linear combination of the previous state estimate x_{k-1}, the system control input u and a random mean process noise w which can be statically modelled. The other parameters are the process transition matrix A and transformation matrix B. Equation 3.2 represents the relationship between the physical measurement or observation z_k and the system state, where H is a transition factor relating the state estimate to the observation and v is the mean measurement or observation noise. This equation represents the *observation model*.

$$x_k = Ax_{k-1} + Bu + w \qquad \{3.1\}$$

$$z_k = Hx_k + v \qquad \{3.2\}$$

State-space representation is fundamental to virtually all linear estimators [39]. The following represents a generic form of the Kalman filter. It is a discretized Kalman filter algorithm [39].

The discrete Kalman filter high level algorithm

Time update phase: predicting the upcoming state	Predict the upcoming (*a priori*) mean system state using the *a posteriori* state estimate from a prior time step: $$\hat{x}_k^- = A\hat{x}_{k-1} + Bu_k$$ Project the *a priori* error covariance: $$P_k^- = AP_{k-1}A^T + Q$$
Observation update phase: correcting the predictions	Compute the Kalman gain: $$K_k = P_k^- H^T (HP_k^- H^T + R)^{-1}$$

> Update the predicted state with the Kalman gain and observation:
>
> $$\hat{x}_k = \hat{x}_k^- + K_k(z_k - H\hat{x}_k^-)$$
>
> Update the error covariance:
>
> $$P_k = (I - K_k H)P_k^-$$

A = state transition matrix

B = transformation matrix

u = control input

Q = process noise covariance matrix

H = state-observation transition matrix

R = observation noise covariance matrix

The *residual,* which is the difference between the actual observation and the predicted observation, i.e. $(z_k - H\hat{x}_k^-)$, shows how far the estimated observation is from the actual observation [39]. In the case of a process containing more than one observed quantity, the Kalman filter's algorithm can be expanded upon, i.e.

$$\hat{x}_k = \hat{x}_k^- + K_{k(i)}\left(z_{k(i)} - H_i\hat{x}_k^-\right) + K_{k(i+1)}\left(z_{k(i+1)} - H_{i+1}\hat{x}_k^-\right) + \;... \qquad \{3.3\}$$

i represents each set of observation. In the case of a multi-observation system e.g. a multi-sensor system, the observation update phase of the Kalman algorithm would be repeated for each set of observations.

The Kalman filter's performance deteriorates significantly when dealing with non-linear systems. A version of the Kalman filter that locally linearizes about the current estimate of a discrete-time non-linear system (as shown in equations 3.4 by the non-linear state transition and state-observation transition a and h respectively) is the *Extended Kalman Filter (EKF)* [39].

$$x_k = a(x_{k-1}, w) \qquad \{3.4a\}$$

$$z_k = h(x_k, v) \qquad \{3.4b\}$$

This linearization occurs locally during a time-step using the linearized transformation matrices a and h, which are approximated as Taylor series sums, generally to the first order (equation 3.5 is an example showing the linear approximation of the state transition matrix). This pseudo-linear transformation renders the system non-Gaussian for extremely non-linear systems. This is because the normal distribution parameters mean and variance are not maintained by the transformation. Depending on the level of non-linearity, its performance could range from reasonably good for low non-linearity to poor for highly non-linear systems.

$$A(x_k) \approx A(x_{k-1}) + \frac{\delta A(x_{k-1})}{\delta x_{k-1}}(x_k - x_{k-1}) \qquad \{3.5\}$$

There is however another variation of the Kalman filter which better maintains the normal distribution through non-linear transformations called the *Unscented Kalman Filter (UKF)* [39] [40]. It works by approximating a Gaussian distribution rather than the transformation function [40]. It was first developed by Julier and Uhlmann [40]. Its computational costs are highly dependent on the complexity of the non-linear transformation and the size of its sample points. It works on non-linear Gaussian systems of the form similar to those of equation 3.4. The UKF employs carefully selected *sample* or *sigma points* around the current state and propagates a new state and covariance from each of the sigma points through the non-linear transition matrices. These new values are then used in estimating the new state and covariance. Where the EKF achieves first order accuracy (Taylor series expansion), the UKF reaches second order accuracy and even reaches third order for Gaussian inputs [41].

There are many other variations of the KF depending on the type of system model and performance requirements such as the *quadrature Kalman filter* and *divided difference filter* (UKF variants), *Gaussian sum Kalman filter* (which can handle multi-peak probability distribution) [21] etc.

3.4.2.2 Particle Filter

The most noticeable property of the aforementioned Kalman filters is that the system representations have to be of the Gaussian distribution form. The *particle filter (PF)* also known as the *Sequential Monte Carlo (SMC)* is a numerical estimator without probability distribution requirements placed on the arbitrary (linear or non-linear) system model. When dealing with non-linear systems, the KF finds an exact solution to a simplified system model while the PF looks to find an approximate solution to the complex, more accurate system model. PF is similar to the UKF in that it makes use of sample points or *particles* to predict a system's state. They however have an extra resampling step where new corrected particles are taken after the state estimate has been updated using the system's observation. The particles are assigned *importance weights* based on their accuracy relative to the estimated state. The new particles are then used at the next time step. Figure 14 is the high level view of a basic PF.

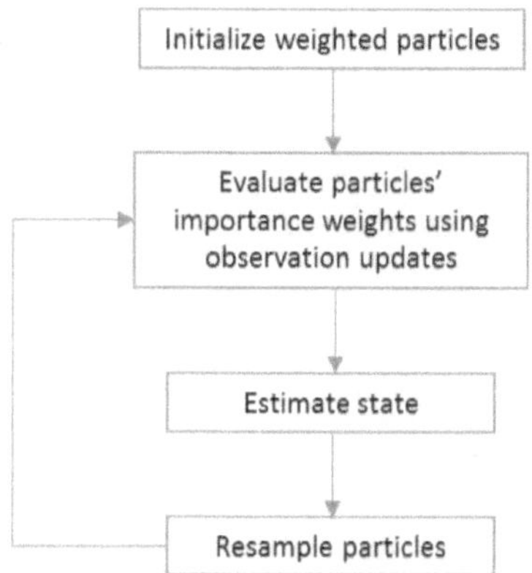

Figure 14: Generic particle filter's high level recursive algorithm

These filters are not exclusive in their functionality as various filters have been created from a combination of different aspects of different filters. For example, Wan and Merwe [41] replaced the generic PF's particle creation technique with the UKF's systematic sample points' creation algorithm creating a filter called the Unscented Particle Filter (UPF). When compared with the performances of the generic EKF, UKF and PF, it was observed that the UPF's performance was superior for a 1D non-linear state estimation and equally superior to the other filters but on par with an EKF-based PF filter for non-linear parameter estimation.

3.4.2.3 Complementary Filter

This filter produces a state estimate from the weighted sum of the state predictions from multiple observations or system models. When using a complementary filter (CF) in navigation and localization, estimates from sensor outputs with high degradation rates in the long term are passed through a high pass filter while outputs susceptible to small disturbances are passed through a low pass filter. The transfer function in equation 3.6 is an example of a CF implementation. The state x is estimated by adding the predicted state x_1 through a 2nd order high pass filter and a second predicted state x_2 through a 2nd order low pass filter.

$$x = \frac{s^2}{s^2 + K_1 s + K_2} x_1 + \frac{K_1 s + K_2}{s^2 + K_1 s + K_2} x_2 \qquad \{3.6\}$$

The CF is simpler and computationally less expensive than the Bayes' filters discussed making it more suited to systems having limited resources. The Bayes' filters however provide a more accurate AHRS estimation as they theoretically minimise the squared estimate error [42].

4 ESTIMATION OF THE SEAHOG'S ATTITUDE

Before an estimation algorithm could be implemented on the iNEMO, a simulation was carried to determine the most suitable method. When deciding on the algorithms to compare, some important factors considered were:

- Field of typical application

- Use prevalence in industry and academia

- Computational size requirement

- Speed of execution

4.1 Simulation of IMU Sensors and AHRS Algorithms

Before implementing an attitude estimating system for the Seahog, an algorithm for this purpose had to be designed. To achieve this, simulation models of a few filters were made and the best performing algorithm was selected. As there were no usable offline sensor data available from the iNEMO, it was decided that simulation models of the sensors (accelerometer and gyroscope) were to be generated.

4.1.1 Simulation Models

The simulation process was broken down into three groups: the system model, the IMU model and the attitude algorithm model.

4.1.1.1 The System Model

The quaternion representation was chosen because of its computational advantages over the Euler notations which have been discussed. As explained in section 3.4.2, the state-space representation of a system such as the Seahog involves two models: the state model and the measurement model. For attitude involving IMUs, it is common to use the measurement from the gyroscope in the state modelling while the other measurements from the accelerometer and magnetometer (when applicable) are employed in the measurement modelling. The Seahog's state was taken to be a seven-element vector made up of the four elements of the quaternion and the gyroscope's biases. The gyroscope's bias was added to the state to account for the possibility of a changing gyroscope bias over time.

The system and measurement models used in the simulations are derived in the following equations. The computations were quaternion-based but the resulting attitude were converted to Euler angles as the results are more user friendly in terms of interpretation in Euler representation.

$$q = \begin{bmatrix} q_0 \\ q_1 \\ q_2 \\ q_3 \end{bmatrix} \qquad \{4.1\}$$

q_0 is the scalar part of quaternion vector q. The angular rate vector relative to the body axes was defined as $\omega = [\omega_x \quad \omega_y \quad \omega_z]^T$.

Using Newtonian formulation, the quaternion was updated at a time instant k after a sampling period dt had passed from the last update:

$$q_k^- = q_{k-1} + \dot{q}_k dt = q_{k-1} + \tfrac{1}{2}\omega q_{k-1}\, dt = (1 + {}^1/_2\,\omega\, dt)q_{k-1} \qquad \{4.2\}$$

The quaternion state transition can therefore be summed up as the following where I is an identity matrix and S_ω is the *skew symmetric matrix* of the angular rate vector about the body axes.

$$q_k^- = \left(I_{4\times4} + \tfrac{1}{2}\begin{bmatrix} 0 & -\omega^T \\ \omega & -S_\omega \end{bmatrix} dt \right) q_{k-1} \qquad \{4.3\}$$

Measurements from the accelerometer were used in the observation model. The accelerometer's measurement prediction y_k^- can be obtained by transforming the normalized gravity vector on the non-rotating reference frame to the body frame using the updated quaternion and the quaternion-based rotation matrix:

$$y_k^- = H(q_k^-)g \qquad \{4.4\}$$

Here, $H(\boldsymbol{q}_k^-)$ is the quaternion-based rotation matrix which transforms the gravity vector $\boldsymbol{g}$ from the reference frame onto the body-fixed frame.

$$\boldsymbol{y}_k^- = \begin{bmatrix} q_0^2 + q_1^2 - q_2^2 - q_3^2 & 2(q_1q_2 + q_0q_3) & 2(q_1q_3 - q_0q_2) \\ 2(q_1q_2 - q_0q_3) & q_0^2 - q_1^2 + q_2^2 - q_3^2 & 2(q_2q_3 + q_0q_1) \\ 2(q_1q_3 + q_0q_2) & 2(q_2q_3 - q_0q_1) & q_0^2 - q_1^2 - q_2^2 + q_3^2 \end{bmatrix} \boldsymbol{g} \qquad \{4.5\}$$

The Euler angles were subsequently recovered from the normalized quaternion using the equivalent relations between the quaternion rotation matrix and the Euler angles-based rotation matrix. The quaternion was normalized to ensure orthogonality since its equivalent Euler rotation matrix is also orthogonal.

$$\phi = \tan^{-1}\left(\frac{2(q_2q_3 + q_0q_1)}{1 - 2(q_1^2 + q_2^2)}\right), \qquad -90\,^\circ < \phi < 90\,^\circ$$

$$\theta = \sin^{-1}(2(q_0q_2 - q_1q_3)), \qquad -90\,^\circ \le \theta \le 90\,^\circ \qquad \{4.6\}$$

$$\psi = \tan^{-1}\left(\frac{2(q_1q_2 + q_0q_3)}{1 - 2(q_2^2 + q_3^2)}\right), \qquad -180\,^\circ \le \psi \le 180\,^\circ$$

The range of each angle was limited by the Seahog's expected range of operation, i.e. a roll or pitch rotation greater than 90 ° would be unlikely. To ensure the full range of the yaw angle is computationally achievable, the *arctan2* functions available in MATLAB and c libraries were used here. The arctan2 function was used for roll none the less.

4.1.1.2 iNEMO's Sensors

A sensor's mathematical model typically includes some transformation function, bias and environmental noises. It has been found that it is better to model each IMU sensor separately for a better model performance [43]. The accelerometer, gyroscope and magnetometer were therefore modelled separately. The following subsections explain each sensor's model.

4.1.1.2.1 Accelerometer's Model

The vectored accelerometer measures the total linear acceleration relative to the body-fixed axis. The iNEMO's accelerometer was calibrated due its inadequate factory state performance. This calibration process is documented in Appendix B. It should be noted that the accelerometer was calibrated at room temperature. The accelerometer has not been calibrated for high or low temperatures. The accelerometer's model was the sum of the linear acceleration a_b of the vehicle, rotated normalized gravity vector $\boldsymbol{g}$, the rotational force effect and sensor measurement noise e_a which was modelled as white Gaussian noise [44]. Since the IMU's frame was aligned with the Seahog's, the axes transformation matrix was not included.

$$z_a = a_b + R_i^b g + (\omega \times \omega \times r_b^{IMU}) + e_a \qquad \{4.7\}$$

The vector r_b^{IMU} represents the direction vector from the body frame to the IMU's frame and R_i^b is the rotation matrix of the inertial to the body frame vector transformation.

4.1.1.2.2 Gyroscope's Model

The gyroscope was modelled as a unit that measures the angular velocities about each body axis. The gyroscope bias ω_b was also included and the measurement noise e_g was modelled as white noise.

$$z_g = \omega + \omega_b + e_g \qquad \{4.8\}$$

The time variant effects on the bias (bias instability) due to temperature change was not taken into account as they were taken to be negligible. This conclusion was reached after the gyroscope's measurement, operating in a stationary position was sampled for over three hours. A temperature change was expected over this period. This temperature change could be attributed mostly to the heat generated by the PCB over time assuming a constant room temperature. The bias change was monitored by splitting the duration up into one hour intervals and computing the mean and standard deviation of each gyroscope axis for all three time intervals. The result of this is shown on Table 8.

Table 8: Gyroscope bias over a three hour period

	x-axis (°/s)		y-axis (°/s)		z-axis (°/s)	
	Mean	Std. dev.	Mean	Std. dev.	Mean	Std. dev.
Hour 1	1.2982	0.8142	-3.4957	0.5156	0.0009	0.0353
Hour 2	1.2966	0.8143	-3.4887	0.5154	0.0010	0.0360
Hour 3	1.2689	0.8158	-3.4667	0.5155	0.0012	0.0395

It can be observed that there was little change in the mean bias value along each axis. Note that the mean values and variances used for noise modelling in each sensor's simulation can be found in appendix C.

4.1.1.3 Models for the Attitude Algorithms

The attitude estimators considered were the Extended Kalman Filter (EKF), Unscented Kalman Filter (UKF) and the Complementary Filter (CF). These filters were considered to be sufficient candidates for attitude estimation. The particle filter was not considered because the advantages offered would not outweigh the added complexity for a fairly simple and predictable system.

4.1.1.3.1 Extended Kalman Filter Model

The EKF model is mostly similar to the standard KF model described earlier in this chapter. The following details the algorithm of the EKF's simulation model. as previously stated, the state vector of both Bayes' filters was a seven-element column vector composed of the quaternion vector and a vector of the gyroscope's mean biases: $\hat{x} = \begin{bmatrix} q \\ \omega_b \end{bmatrix}$.

The discrete Extended Kalman filter high level equations

Time update phase: predicting the upcoming state	Predict the *a priori* system state (mean) at instant t using *a posteriori* state estimate from the previous time step: $$\hat{x}_t^- = F(\hat{x}_{t-1})$$

Project the *a priori* error covariance where Q is the process noise and A is the state Jacobian:

$$\boldsymbol{P}_t^- = A\boldsymbol{P}_{t-1}A^T + Q$$

Observation update phase: correcting the predictions	Predict the accelerometer's measurement: $$\hat{\boldsymbol{y}}_t^- = H(\hat{\boldsymbol{x}}_t^-)$$ Compute the Kalman gain where R is the measurement noise and H_a is the measurement's Jacobian: $$K_a = \frac{\boldsymbol{P}_t^- H_a{}^T}{H_a \boldsymbol{P}_t^- H_a{}^T + R}$$ Update the predicted state with the Kalman gain and measurement (where $\hat{\boldsymbol{y}}_t$ is the actual measurement at time t): $$\hat{\boldsymbol{x}}_t = \hat{\boldsymbol{x}}_t^- + K_a(\hat{\boldsymbol{y}}_t - \hat{\boldsymbol{y}}_t^-)$$ Update the error covariance: $$\boldsymbol{P}_t = (I - K_a H)\boldsymbol{P}_t^-$$

4.1.1.3.2 Unscented Kalman Filter Models

The following describes the standard recursive UKF simulation model.

The discrete Unscented Kalman filter high level equations [41]

UKF parameter initialization	Initialize the sigma points' scaling parameters (where n is the size of the column state vector): $$\lambda = \alpha^2(n + \kappa) - n$$ Initialize the weights of each sigma point's mean and covariance (where X_i is i^{th} sigma point):

$$w_0^m = \frac{\lambda}{n + \lambda}$$

$$w_0^c = \frac{\lambda}{n + \lambda} + 1 + \beta - \alpha^2$$

$$w_i^m = w_i^c = \frac{1}{2(n + \lambda)}, \quad i = 1, \dots, 2n$$

Initialize the state vector and the error covariance matrix:

$$\hat{x} = x_0$$

$$P_x = P_0$$

Time update phase: predicting the a priori state	Generate $2n + 1$ sigma points at time instant $t - 1$ (where $v = n + \lambda$): $$X_0^{t-1} = \hat{x}_{t-1}$$ $$X_i^{t-1} = X_0^{t-1} + \left(\sqrt{vP_x^{t-1}}\right)_i, \quad i = 1, \dots, n$$ $$X_i^{t-1} = X_0^{t-1} - \left(\sqrt{vP_x^{t-1}}\right)_{i-n}, \quad i = n+1, \dots, 2n$$ Predict the *a priori* state for each sigma point using the previous *a posteriori* sigma points by performing an *unscented transformation*: $$X_{i(t)}^- = F(X_i^{t-1}), \quad i = 0, \dots, 2n$$ Calculate the mean predicted state from the weighted transformed points: $$\hat{x}_t^- = \sum_{i=0}^{2n} X_{i(t)}^- w_i^m$$ Project the *a priori* error covariance:

$$P^-_{x(t)} = \sum_{i=0}^{2n} \left(X^-_{i(t)} - \hat{x}^-_t\right)\left(X^-_{i(t)} - \hat{x}^-_t\right)^T w^c_i + Q$$

Update the sigma points by adding the zero mean process noise:

$$X^t_i = X^-_{i(t)} \ , \quad i = 0$$

$$X^t_i = X^-_{i(t)} + \left(\sqrt{(n+\lambda)Q}\right)_i \ , \quad i = 1, \dots, n$$

$$X^t_i = X^-_{i(t)} - \left(\sqrt{(n+\lambda)Q}\right)_{i-n} , \quad i = n+1, \dots, 2n$$

Predict the measurement using each updated sigma point:

$$Y^t_i = H(X^t_i), \qquad i = 0, \dots, 2n$$

Determine the weighted mean from the predicted measurements:

$$\hat{y}^-_t = \sum_{i=0}^{2n} Y^t_i w^m_i$$

Calculate the measurement covariance:

$$P^t_{yy} = \sum_{i=0}^{2n} (Y^t_i - \hat{y}^-_t)(Y^t_i - \hat{y}^-_t)^T w^c_i + R_a$$

Calculate the measurement cross covariance:

$$P^t_{xy} = \sum_{i=0}^{2n} (X^t_i - \hat{x}^-_t)(Y^t_i - \hat{y}^-_t)^T w^c_i$$

Calculate the Kalman gain:

$$K_a = \frac{P^t_{xy}}{P^t_{yy}}$$

Update the predicted state with the Kalman gain and observation (where $\hat{y}_t$ is the actual measurement):

$$\hat{x}_t = \hat{x}_t^- + K_a(\hat{y}_t - \hat{y}_t^-)$$

Update the error covariance:

$$P_x^t = P_{x(t)}^- - K_a P_{yy}^t K_a^T$$

For the UKF, the scaling constants α and κ are generally set such that $1 \gg \alpha \gg 10^{-4}$ and $\kappa \geq 0$ [41]. These parameters determine how far the sigma points are from the mean. For Gaussian systems, β which is a function of the kurtosis (the measure of the heaviness of the tail/outliers of a probability distribution) is set to two for optimal performance [40]. These parameters were set such that the mean and covariance were captured while errors due to the third and fourth (kurtosis) order moments of the distribution were minimised. The aforementioned parameters are relevant to the standard *symmetric set* of $2n + 1$ sigma points [41]. This set is symmetric because $2n$ points are symmetrically distributed about a mean state (as shown in the sigma point generation phase on the previous table). However, a minimum *asymmetric simplex set* of $n + 2$ can be used to capture the mean and covariance of a system's state [41] [40]. The simplex set has the advantage of lower computation cost as computational cost is proportional to the size of the sigma points set. There are several point selection and weight assignment methods that have been suggested for generating a simplex set: with their initial simplex set proposal, Julier and Uhlmann [45] proposed a method that minimized the effects of the distribution's skew by capturing its third central moment (achieved by making the distribution symmetric about its mean); a more popular approach was subsequently proposed by Julier [46] that produced a *spherical simplex* set whose points fall on a hypersphere. This was observed to be more numerically stable than the initial *minimal skew* set, especially when dealing with high dimension states. An even more reduced set of $n + 1$ was proposed by Cheng and Liu [47] which produced reasonably accurate signal tracking estimates when compared with the standard UKF and the EKF. For this project though, the spherical simplex (SS) set was added to the list of filters to be tested as it would provide a cheaper alternative to the standard UKF in terms of computational implementation if its performance is sufficient. The operational difference between the standard UKF and the SS-UKF are their sigma point generation phases. The following is the point generation algorithm of the SS-UKF with its accompanying parameters.

The Spherical Simplex UKF set generation algorithm

Initialization	Initialize the sigma points' weights (where n is the size of the state vector), $0 < W_0 < 1$:

$$W_1 = {(1 - W_0)}\big/{(n + 1)}$$

$$w_0 = 1 + {(W_0 - 1)}\big/{\alpha^2}$$

$$w_1 = {W_1}\big/{\alpha^2}$$

Initialize the weights of each sigma point's mean and covariance:

$$w_0^m = w_0$$

$$w_0^c = w_0 + 1 + \beta - \alpha^2$$

$$w_i^m = w_i^c = w_1, \qquad i = 1, \dots, n + 1$$

Initialize the first row elements of the error covariance's scaling factor matrix $\mathbf{Z}$, which is a 7×9 matrix; $\mathbf{Z}_i$ is an i-nth column vector:

$$\mathbf{Z}_{0,3-8} = 0; \quad \mathbf{Z}_1 = {-1}\big/{\sqrt{2w_1}}; \quad \mathbf{Z}_2 = {1}\big/{\sqrt{2w_1}}$$

The j-nth row (from the second row, i.e. $j = 1$) and i-nth column of the sigma points weight matrix $\mathbf{Z}$:

$$Z_i = \begin{cases} [\mathbf{0}_{6\times 1}], & for\ i = 0 \\[2mm] \begin{bmatrix} \mathbf{0}_{(i-3)\times 1} \\ (i-1)\big/\sqrt{i(i-1)w_1} \end{bmatrix}, & for\ i = 3, \dots, n+1 \\[4mm] -1\big/\sqrt{j(j+1)w_1}, & for\ i = 1, \dots, n \end{cases}$$

The fully defined Z matrix can be found in

Appendix C

Time update phase: sigma point generation	Generate $n+2$ sigma points at time instant $t-1$: $X_0^{t-1} = \hat{\boldsymbol{x}}_{t-1}$ $X_i^{t-1} = X_0^{t-1} + \sqrt{P_x^{t-1}}Z_i\,, \qquad i = 0, \dots, n+1$

When taking the square root of a non-diagonal matrix, in the context of the UKF, Rhudy et al. [48] found that the best technique out of eight square root methods: three analytical and five iterative, was the *Cholesky decomposition*. This conclusion was based on error minimization and computational execution duration performances. The *lower triangular Cholesky decomposition* was therefore used in the simulations when dealing with non-diagonal matrices.

For both the EKF and UKF models, the process covariance Q is a diagonal square matrix composed of the quaternion covariance diagonal matrix $\boldsymbol{Q}_q$ and the gyroscope bias covariance diagonal matrix $\boldsymbol{Q}_b$. The non-diagonal elements were taken to be zero as cross-correlation between the gyroscope's measurements and the quaternion's parameters were taken to be negligible.

$$Q = \begin{bmatrix} \boldsymbol{Q}_q & \mathbf{0}_{4\times 3} \\ \mathbf{0}_{3\times 4} & \boldsymbol{Q}_b \end{bmatrix} \tag{4.9}$$

4.1.1.3.3 Complementary Filter Model

The equation of the traditional CF model employs second order filters on the accelerometer and gyroscope estimates. The accelerometer's tilt angle estimation is passed through a low pass filter to remove short term high frequency noise while the gyroscope's integrated data is passed through a high pass filter to remove its long term drift. The following transfer equation shows this:

$$\boldsymbol{\theta} = \frac{s^2}{s^2+K_1 s+K_2}\left(\frac{1}{s}\dot{\boldsymbol{\theta}}_g\right) + \frac{K_1 s+K_2}{s^2+K_1 s+K_2}\boldsymbol{\theta}_a \qquad \{4.10\}$$

This representation however has flaws such as the potential for numerical instability caused by gimbal lock, the decoupling of the Euler angles i.e. each axis of rotation requires its own sensor data set in order to be able to estimate each respective attitude. This shortcoming made the quaternion representation preferable. Valenti et al. [49] proposed a quaternion-based complementary filter. The algorithm for this filter is as follows.

The discrete quaternion-based Complementary filter equations

Time update phase: predicting the upcoming state	Predict the state quaternion at instant t using the measurement from the gyroscope (where $\boldsymbol{I}$ is an identity matrix and dt is the sampling period): $$q_t^- = \left(I + \frac{1}{2}\begin{bmatrix} 0 & -\boldsymbol{\omega}^T \\ \boldsymbol{\omega} & -S_\omega \end{bmatrix} dt\right) q_{t-1}$$
Observation update phase: correcting the prediction	Predict a gravity vector through the reverse rotation of the acceleration's vector by quaternion q_t^{-*} which is the conjugate of q_t^-: $$\boldsymbol{g}^- = R(q_t^{-*})\boldsymbol{a}_t$$ Determine the rotation error $\boldsymbol{\Delta q}$ between the gravity vector and the normalised predicted gravity vector $\hat{\boldsymbol{g}}^-$ by solving: $$\hat{\boldsymbol{g}}^- = R(\boldsymbol{\Delta q})\boldsymbol{g}$$ Due to the accelerometer's high frequency noise effects on the delta quaternion $\boldsymbol{\Delta q}$, interpolate

between the predicted gravity vector and actual gravity vector using quaternion representation (K is a gain characterising the filter's cut-off frequency). *LERP* (Linear Interpolation) if the angle between the gravity vectors is small, i.e. α < 20 °, otherwise *SLERP* (Spherical Linear Interpolation) [46]:

$$\text{LERP: } \Delta\hat{\mathbf{q}} = (1 - K)\mathbf{q}_I + K\Delta\mathbf{q}$$

$$\text{SLERP: } \Delta\hat{\mathbf{q}} = \frac{\sin((1-K)\alpha)}{\sin\alpha}\mathbf{q}_I + \frac{\sin(K\alpha)}{\sin\alpha}\Delta\mathbf{q}$$

Update the quaternion through multiplication. Normalise $\mathbf{q}_t$ after the update:

$$\mathbf{q}_t = \mathbf{q}_t^- \otimes \Delta\hat{\mathbf{q}}$$

Note that any filter constant not quantified in this section can be found in appendix C.

4.2 Algorithm Simulation Results

Attitude estimates for underwater UUVs are typically accurate to within the following ranges [1]:

Roll, Pitch	< 1 °
Yaw	< 2 °

The filters were put through several test scenarios of which most were in mechanical equilibrium. The Root Mean Square Error (RMSE) was used as an evaluation metric for analysing the filters' performances over a specific period. RMSE is the average deviation from the ideal or expected. RMSE also provides the added advantage of having the same unit as the Euler angles which allows for a more intuitive analysis with regards to accuracy levels.

$$RMSE = \sqrt{\frac{1}{N}\sum_{k=1}^{N}(\hat{x}_k - x_k)^2} \qquad \{4.11\}$$

N is the sample size, $\hat{x}_k$ is the estimated value and x_k is the expected value at instant k. Similar to the RMSE is the Mean Absolute Error (MAE) which also provides an average error. The RMSE was preferred because where the MAE gives the same weight to each error, large errors more significantly affect the RMSE values due to the squaring of the error. This ensures the amplification of instances of poor performance by any of the filters.

Test scenarios were achieved by generating a known attitude vector at every time instant, passing it through the gyroscope and accelerometer's simulation models which both then produced distorted sensor data that was used by each filter. Each filter used the same data set for each scenario to ensure similar test conditions across all filters. Each filter simulation was sampled at the standard Seahog rate of 50 Hz (as underwater vehicles are typically low rate systems) and run for a duration of 200 seconds.

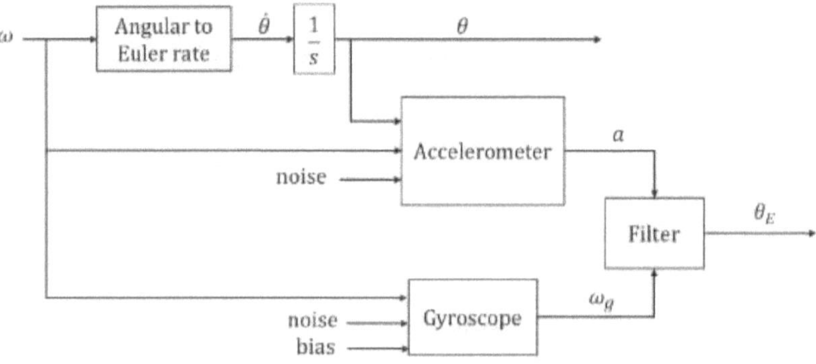

Figure 15: Simulation flowchart for each algorithm

Figure 15 details the overall simulation process followed in the following simulations. Here, θ_E represents the estimated angle and θ is the actual angular position. It summarises the sensor models previously explained in detail and the subsequent filter models by showing the inputs and outputs of each sensor and process. The following subsections detail each scenario tested and the ensuing results.

4.2.1 No Filter

Firstly, before getting into the various algorithm tests, a simple base motion simulation was run without correction. The IMU's quaternion was updated based on the gyroscope's measurements only as shown in equation 4.12.

$$q_k^- = \left(I_{4\times4} + \frac{1}{2}\begin{bmatrix} 0 & -\omega^T \\ \omega & -S_\omega \end{bmatrix} dt\right) q_{k-1} \qquad \{4.12\}$$

The purpose of this was to show why an attitude algorithm is necessary as the estimates were expected to be susceptible to high frequency noise and long-term drift. The following is the result of applying a sinusoidal motion in the pitch direction.

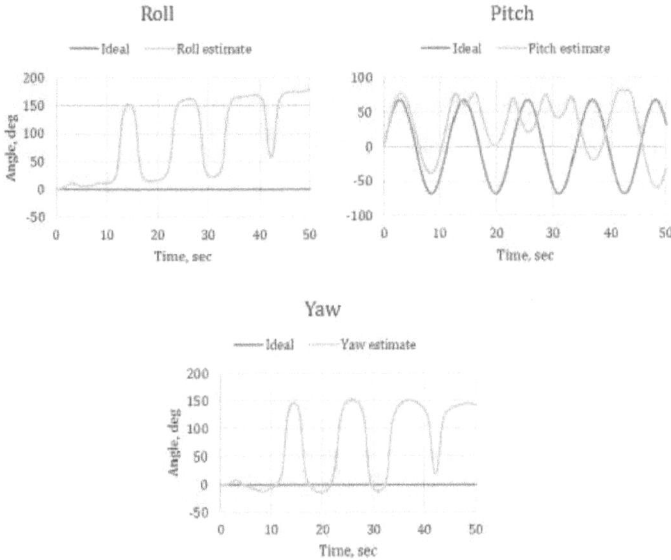

Figure 16: Results from the no filter simulation

From the graphs, it was observed that in the long term, all the estimates were divergent. This served to highlight the necessity of a sensor fusion filter. In each subsequent test scenario, an unfiltered estimation was performed. This would serve to aid the filters' performance analyses.

4.2.2 Base Motion (Scenario 1)

This testing scenario was such that a sinusoidal was applied along the pitch angle direction. The following figure shows this motion and the estimation error for each filter at each sample time instant.

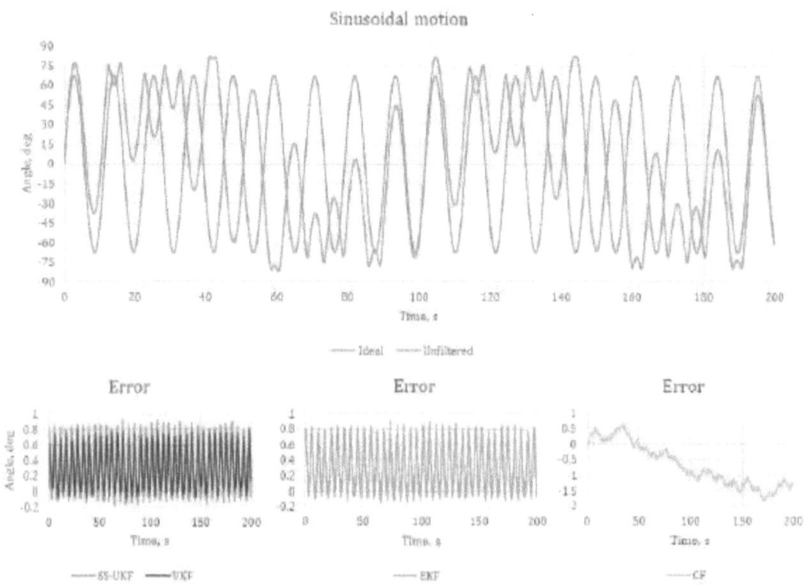

Figure 17: Pitch sinusoidal motion simulation results

The following were the resulting RMSE values from each filter's simulation run.

Table 9: Scenario 1 RMSE results

Filter	RMSE
EKF	0.4160 °
UKF	0.4000 °
SS-UKF	0.4171 °
CF	0.9655 °

4.2.3 Coupled Sinusoidal Motion (Scenario 2)

In this scenario, small-amplitude-sinusoidal motions were generated in the roll and pitch directions. This naturally led to a response in the yaw direction as the axes are coupled.

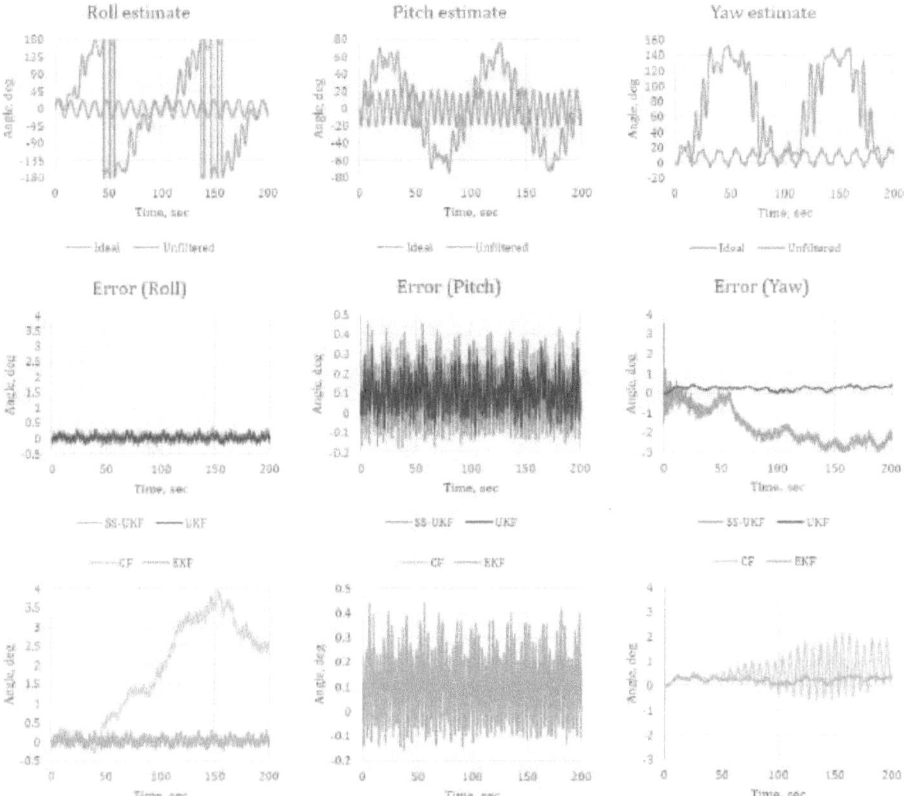

Figure 18: Coupled sinusoidal motion simulation results

The following were the resulting RMSE values from each filter's simulation run.

Table 10: Scenario 2 RMSE results

Filter	Roll	Pitch	Yaw
EKF	0.0897 °	0.1440 °	0.2602 °

UKF	0.0796 °	0.1319 °	0.2608 °
SS-UKF	0.0952 °	0.1472 °	1.9138 °
CF	2.2301 °	0.4016 °	0.8138 °

From the table, it can be observed that all the filters were able to provide relatively low estimation errors due to their respective low RMSEs except for the CF's roll and yaw estimates and the SS-UKF's yaw estimate which appeared to diverge from the ideal.

4.2.4 Instantaneous Step Motion (Scenario 3)

This scenario involved applying an almost instantaneous change in angular position in the pitch direction as shown below.

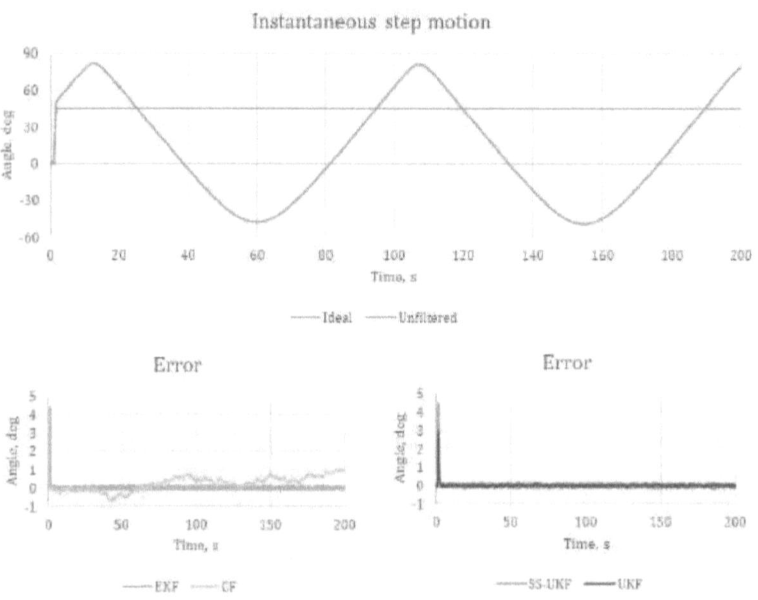

Figure 19: Instantaneous motion simulation responses

The following were the resulting RMSE values from each filter's simulation run.

Table 11: Scenario 3 RMSE results

Filter	RMSE
EKF	0.2108 °
UKF	0.1468 °
SS-UKF	0.2153 °
CF	0.4572 °

4.2.5 Gradual Step Motion (Scenario 4)

Here, a pitch motion of constant angular velocity from 0 ° to 30 ° was implemented.

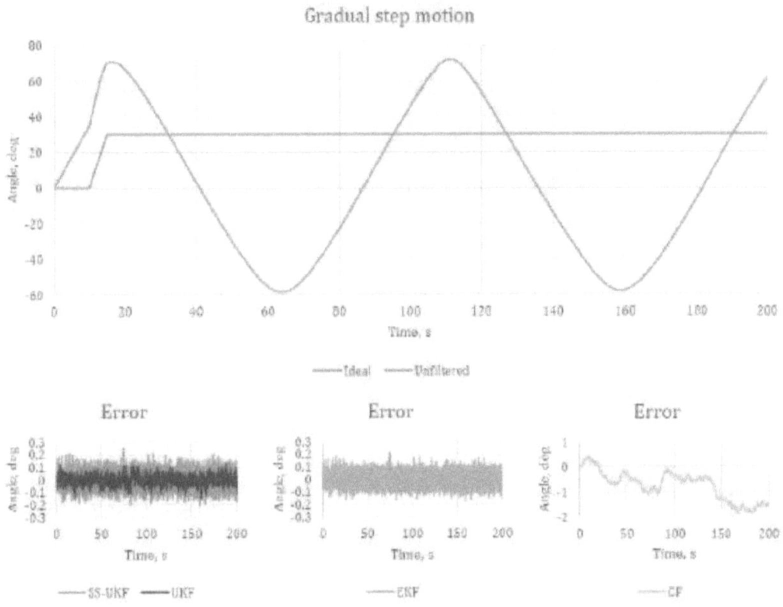

Figure 20: Gradual step motion simulation results

The following were the resulting RMSE values from each filter's simulation run.

Table 12: Scenario 4 RMSE results

Filter	RMSE
EKF	0.0494 °
UKF	0.0378 °
SS-UKF	0.0581 °
CF	0.9442 °

This scenario produced the most accurate attitude tracking performance by all filters barring the CF as evidenced by their small RMSE values.

4.2.6 Yaw Motion (Scenario 5)

A sinusoidal motion in the yaw direction was simulated under this scenario.

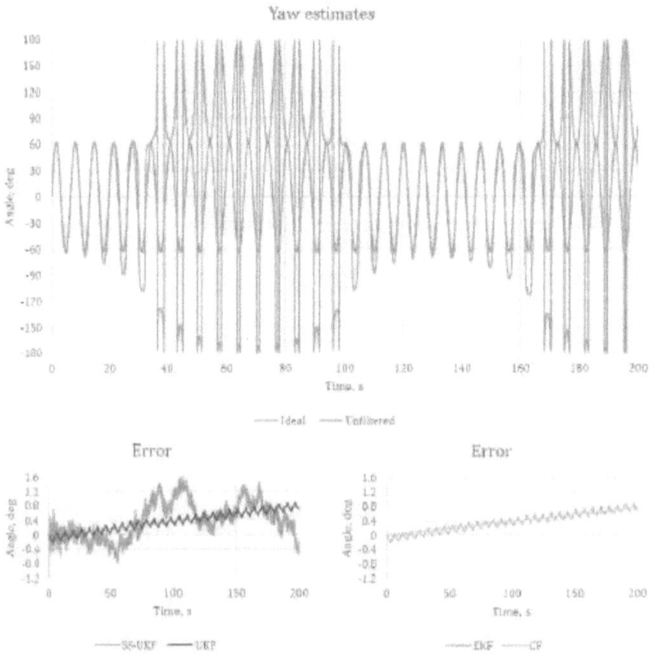

Figure 21: Sinusoidal yaw motion simulation responses

The following were the resulting RMSE values from each filter's simulation run.

Table 13: Scenario 5 RMSE results

Filter	RMSE
EKF	0.4506 °
UKF	0.4523 °
SS-UKF	0.6426 °
CF	0.4488 °

4.2.7 Linear Acceleration (Scenario 6)

A constant linear acceleration of 0.5 m/s^2 in the vehicle's surge direction was applied for the entire duration of the simulations in this scenario. This scenario facilitated in the filters' performance analysis for a vehicle experiencing a constant force. While the vehicle is accelerating, a gradual step motion similar to scenario 4 was applied in the pitch direction.

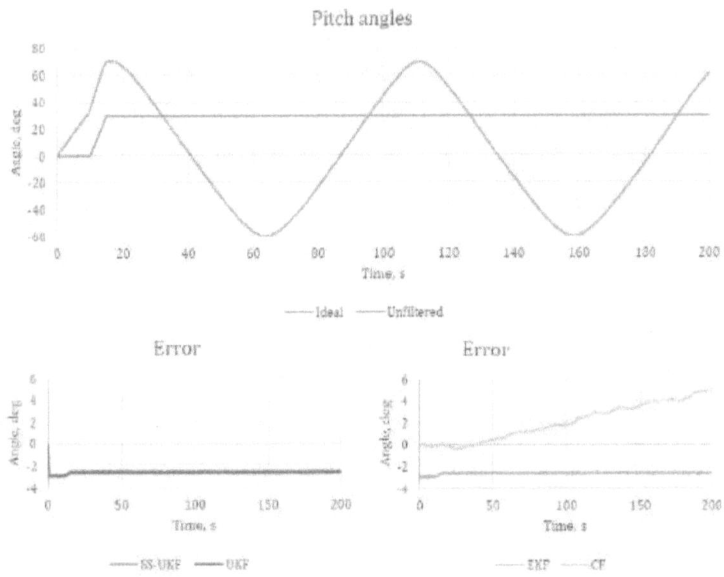

Figure 22: The pitch responses from the simulations

The following were the resulting RMSE values from each filter's simulation run.

Table 14: Scenario 6 RMSE results

Filter	RMSE
EKF	2.6165 °
UKF	2.6120 °
SS-UKF	2.6576 °
CF	2.6167 °

This scenario produced similar responses from the Bayesian and complementary filters as evident from the similar RMSE values.

4.2.8 Excessively Noisy Sensors Simulation

The estimation performance of a filter was tested under severe measurement distortion conditions. The noise source could be due to ocean turbulence or the vibrations due to the vehicle's dynamic thrust outputs. The simulations were based on the EKF. Significant Gaussian noise of more than ten times the magnitude of the measurement noise variance was added to the measurement simulation. This was added to both the angular rate and acceleration measurements but on separate EKF executions. The following is the EKF's estimates while noisy angular rate measurements were observed.

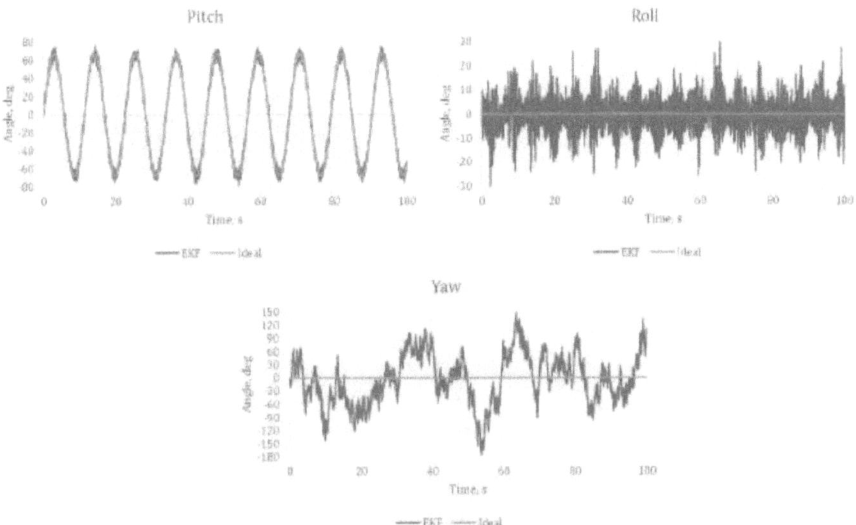

Figure 23: EKF's Euler angle estimates under excessive angular rate measurement noise

The following were the results from the EKF simulations involving noisy acceleration measurements.

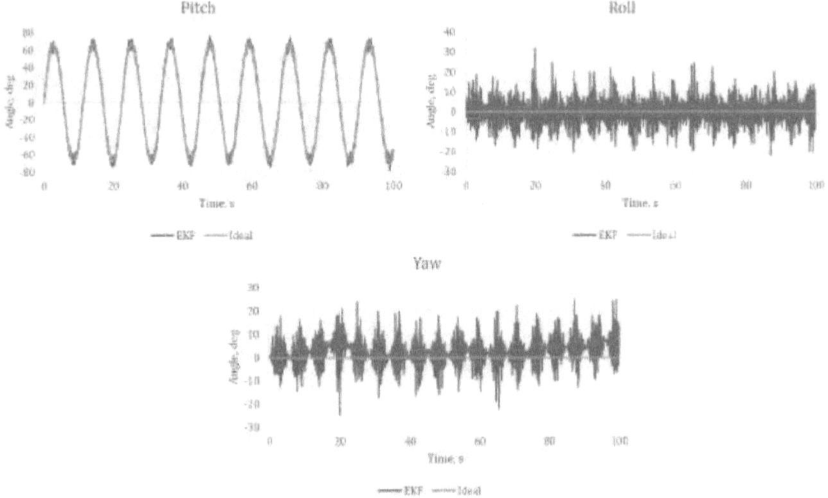

Figure 24: EKF's Euler angle estimates under excessive acceleration measurement noise

4.2.9 Filter Loop Execution Duration

The duration for a single loop of each filter to compute on MATLAB was investigated. It should be noted however that the results from this test could not be directly used in estimating how long each loop cycle would run on the STM32F1 microprocessor as the code would be compiled and executed differently by different processors. CPU operations were not well understood enough for deductions to be made so this subsection only serves to inform on how each filter performs relative to the others with regards to computational speed. Considering that a sample frequency of 50 Hz was used in the simulations, the duration of a cycle of an algorithm should be well under 20 ms. For each filter, the duration of each iteration was recorded and their average was calculated. The following table shows the average time taken for each filter to execute one cycle of its algorithm.

Table 15: One cycle execution period of each algorithm

Filter	MATLAB's simulation average cycle duration
EKF	$62.0 \times 10^{-6} \mu s$
UKF	$283 \times 10^{-6} \mu s$
SS-UKF	$176 \times 10^{-6} \mu s$
CF	$24.8 \times 10^{-6} \mu s$

As expected, the UKF was the most computationally taxing of the tested filters due to its use of multiple high dimensional matrices. The CF executed over the shortest period as it was the shortest algorithm to implement.

4.2.10 Results Analysis

The following is a summary of the scenario tests. The highlighted (green) values represent the best performance in each scenario which was determined by the least RMSE value.

Table 16: Summary of RMSE results from all test scenarios

Filter	1	2 (average)	3	4	5	6
EKF	0.4160 °	0.1858 °	0.2108 °	0.0494 °	0.4506 °	2.6165 °
UKF	0.4000 °	0.1792 °	0.1468 °	0.0378 °	0.4523 °	2.6120 °
SS-UKF	0.4171 °	0.2770 °	0.2153 °	0.0581 °	0.6426 °	2.6576 °
CF	0.9655 °	0.7252 °	0.4572 °	0.9442 °	0.4488 °	2.6167 °

From Table 16, it is obvious that the UKF generally provided a slightly better performance due to it having the least RMSE values. It was however not significantly better than its other Bayesian counterparts. Some factors that could affect the relative performances of the EKF and UKF thereby minimizing the UKF's advantages over the EKF are:

- If the algorithm is executed at a high enough sampling frequency, the EKF's linearization error gets minimized making its estimate more accurate [50].
- The UKF's advantages are equally diminished if the system is not particularly truly non-linear, i.e. the nonlinearities are small.
- LaViola also suggested that because of the low variances associated with the attitude system, particularly with the sensor and system models' variances being in the region of 10^{-4}– 10^{-6}, this could minimise the 4th order moments thereby diminishing the UKF's 4th order accuracy effects [50].

With all these considered, the EKF was deemed a sufficient algorithm for attitude estimation to be implemented on the Seahog. From the summary Table 16, the steady-state EKF tilt angle estimate RMSE all fell within the ideal range of < 1 °. Computational speed was also a major consideration as an EKF cycle was only slower than the CF according to the Table 15 test results.

Looking at Figure 23 and Figure 24 which were the results from EKF simulation with excessive noise added to the angular rate and acceleration measurements respectively, inaccurate roll and yaw estimates were observed while a pitch angle with high-amplitude noise was estimated. These performances were to be expected as the measurement's excess noise variances were significantly greater than the value expected by the EKF, causing a deterioration in the estimates.

In certain test scenarios, some of the Bayesian filters' estimate errors were sinusoidal in nature. This was the case for the EKF, UKF and SS-UKF in scenario 1 (Figure 17) and the EKF and UKF in scenario 5 (Figure 21). This implies that the error under those testing conditions could be attenuated due to their predictable nature.

4.2.10.1 Yaw Estimation

The EKF, UKF, SS-UKF and CF were expected to estimate the tilt angles (roll and pitch) more accurately than the yaw because of the absence of measurements that could be used to correct the system model's yaw prediction. The tilt angles on the other hand were corrected using the accelerometer's measurements. However, the more accurately a gyroscope is characterized, the better the angular rate compensation which ultimately results in less error being introduced into the yaw estimation. This characterization in addition to including the standard bias and sensor noise would include a time-variant drift.

During a no motion state, a yaw drift was expected due to the error accumulation in the yaw estimate over time. From the gyroscope's data collected and discussed in section 4.1.1.2.2, after a three hour sampling, it was observed that the z-axis gyroscope's bias was within the range of $-1\ °/s$ to $0\ °/s$. To compensate for the bias at any given time instant, any z-axis gyroscope measurement within the range $-1\ °/s \leq \omega_z \leq 0\ °/s$ was considered to be a zero rotation, i.e. that particular z-axis gyroscope measurement was effectively set to zero. This compensation method would have been ideal assuming rotation measurement about the z-axis was ordinarily added to the bias. This however potentially creates a dead band between -1 °/s to 0 °/s. The following graph display the yaw estimate from simulations for two cases: the normal EKF implementation and the EKF implementation with the z-axis gyroscope dead band.

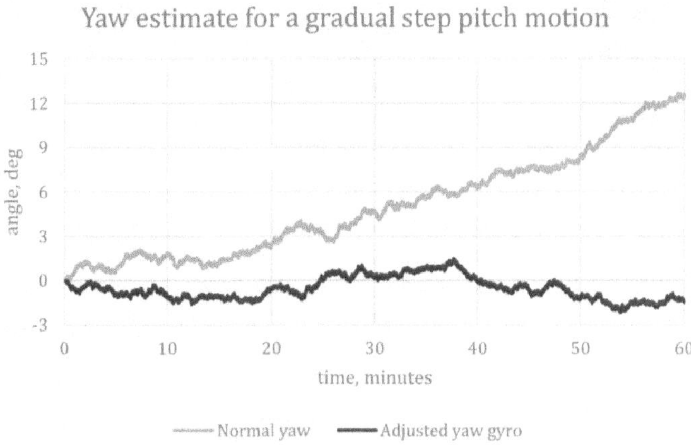

Figure 25: Yaw estimates during zero motion in yaw direction

Figure 25 shows the yaw estimates from a gradual pitch step (0 ° to 30 °) motion while there was no motion in the roll and yaw directions. This simulation was run for a time period over an hour to observe the yaw drift. It is evident that the normal yaw estimate drifted at about a pseudo-linear rate of about 12 °/hr. The EKF simulation with the z-axis gyroscope measurement limit imposed produced a yaw estimate with little drift over time.

4.3 EKF's Implementation on the iNEMO

The EKF algorithm was subsequently implemented on the iNEMO. Some tests were performed to observe the algorithm's performance. The testing rig employed was a three-axis gimbal with a calibrated potentiometer attached to measure the yaw angle. This potentiometer was subsequently used as a reference point for the iNEMO's yaw estimate.

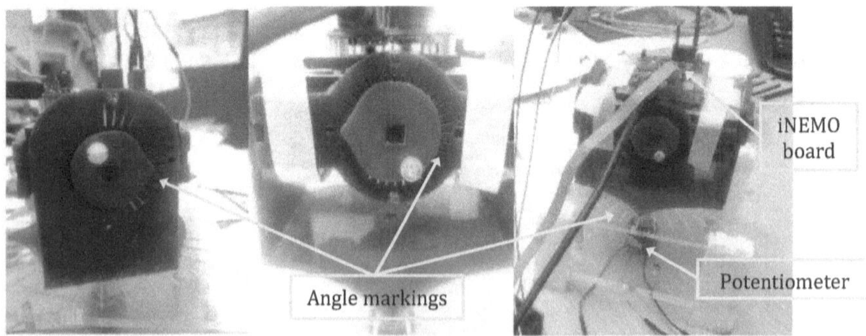

Figure 26: Gimbal used in testing iNEMO's attitude outputs

The EKF was executed on the STM32F1 microprocessor on the iNEMO. The duration a cycle of the EKF took was measured from the duration an LED on the iNEMO stayed on. The LED was put on at the start of a cycle and put off at the end. It was found that a typical EKF cycle took about 8 ms to execute, which was about one third of the sampling period.

4.3.1 Yaw Estimation

Firstly, a basic EKF was implemented on the iNEMO. Its yaw estimate was sampled for over a minute while the iNEMO was stationary. Below is a summary of how it was implemented on the iNEMO. $\omega_g - \omega_b$ is the gyroscope's measurement minus the mean bias.

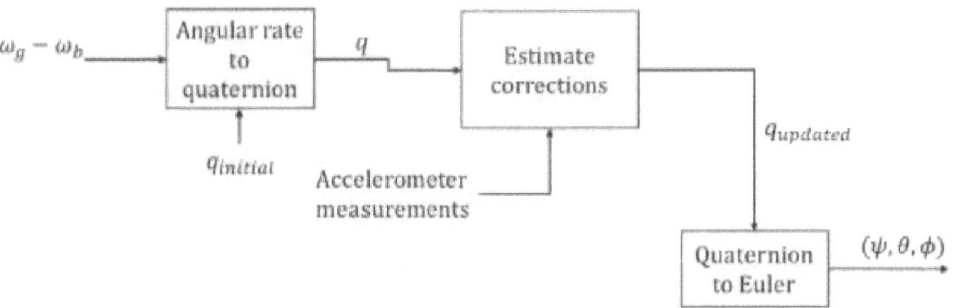

Figure 27: iNEMO's first EKF implementation

The following is a result from the iNEMO's yaw estimate for a stationary state. It was decided to run two versions of the EKF algorithm based on sensor data input: usage of all accelerometer and gyroscope data and usage of only the z-axis gyroscope data. In the second case, the other angular rate inputs were set to zero while the gravity vector was assumed to be parallel to the iNEMO's z-axis.

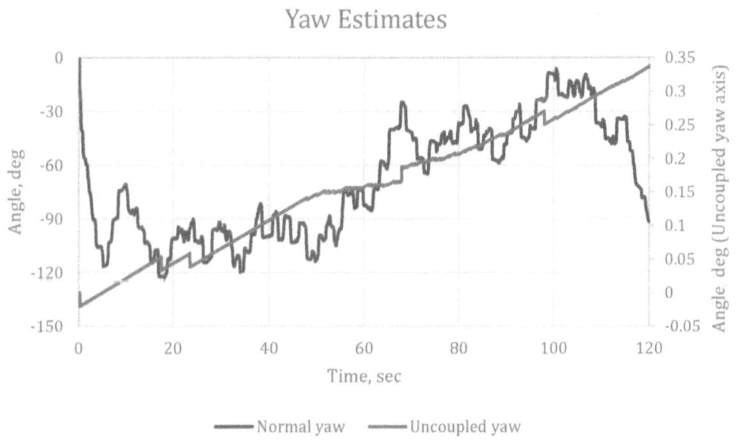

Figure 28: Yaw estimates from iNEMO while stationary

The normal yaw estimate (featuring all sensor data; left axis) grew at a high rate and in seemingly random directions. This can be attributed to the sensors' data all having some inherent random error which in turn meant noisy roll and pitch estimates. The yaw estimate from the EKF filter was quite sensitive to the roll and pitch rates since all three angles are coupled. On the other hand, the uncoupled yaw estimate (right axis of Figure 28) was pseudo-linear since its only sensor data is the z-axis' angular rate. It has a linearized growth rate of about 10.5 °/hr. Isolating the z-axis angular rate data is however not suitable. Coupled angular motion about the iNEMO's x and y axes will produce a yaw angular displacement, which will be undetectable by this approach.

Considering the normal approach, the yaw estimate should not be affected by the correction phase's effects on the quaternion. This therefore led to a different implementation that takes this into account was considered. This EKF implementation and ensuing result are as follows.

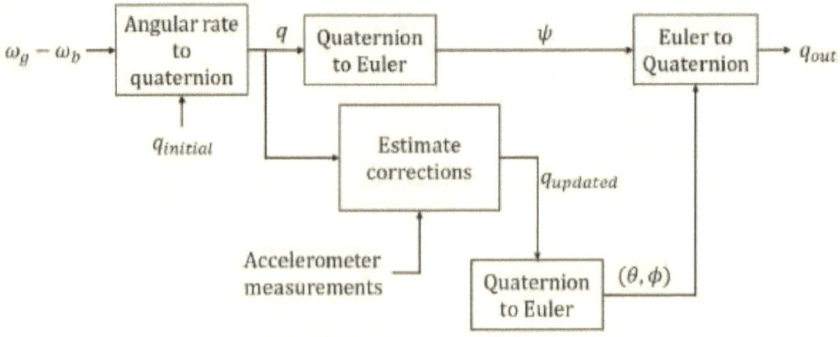

Figure 29: iNEMO's EKF second implementation

Note that for the next EKF cycle, $q_{initial} = q_{out}$. The equations used for the conversion of the Euler angles to quaternion applicable to the rotation sequence $R(\phi)R(\theta)R(\psi)$ were [51]:

$$r = \frac{\phi}{2}, p = \frac{\theta}{2}, y = \frac{\psi}{2} \qquad \{4.13\}$$

$$q_0 = \cos r \cos p \cos y + \sin r \sin p \sin y$$

$$q_1 = \sin r \cos p \cos y - \cos r \sin p \sin y \qquad \{4.14\}$$

$$q_2 = \cos r \sin p \cos y + \sin r \cos p \sin y$$

$$q_3 = \cos r \cos p \sin y - \sin r \sin p \cos y$$

Quaternion to Euler equations:

$$\phi = \tan^{-1}\left(\frac{2(q_2 q_3 + q_0 q_1)}{1 - 2(q_1^2 + q_2^2)}\right)$$

$$\theta = \sin^{-1}(2(q_0 q_2 - q_1 q_3)) \qquad \{4.15\}$$

$$\psi = \tan^{-1}\left(\frac{2(q_1 q_2 + q_0 q_3)}{1 - 2(q_2^2 + q_3^2)}\right)$$

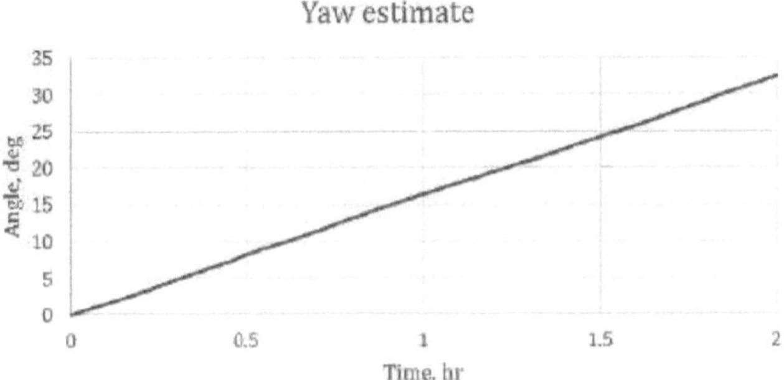

Figure 30: Stationary yaw estimate from the iNEMO after implementing the second version of the EKF

The yaw angle was estimated without creating a gyroscope z-axis dead-band as originally suggested in section 4.2.10.1. From the above figure, where the iNEMO was left stationary for more than two hours, a zero drift rate of about 16.20 °/hr was observed. The difference between these two implementations is that the yaw was affected by the tilt correction in the first case but this effect had been ignored in the second case by estimating the yaw before correcting the quaternion using accelerometer measurements. The second implementation was therefore applied in the subsequent tests. Since the yaw drift rate (16.2 °/hr – which is about 5% of the yaw range after an hour) was viewed to be relatively small, the dead-band (–1 °/s →

0 °/s) on the z-axis gyroscope suggested in section 4.2.10.1 was not included in the iNEMO's EKF implementation.

More yaw estimation were performed by recording the iNEMO's yaw estimates and estimates from the calibrated potentiometer shown in the Figure 26. The following two graphs were estimates of pure yaw rotations (no roll, pitch rotation). The iNEMO's estimate is compared with the potentiometer's estimate which was assumed to be relatively accurate as it had been previously calibrated to be in alignment with the gimbal's angle markings.

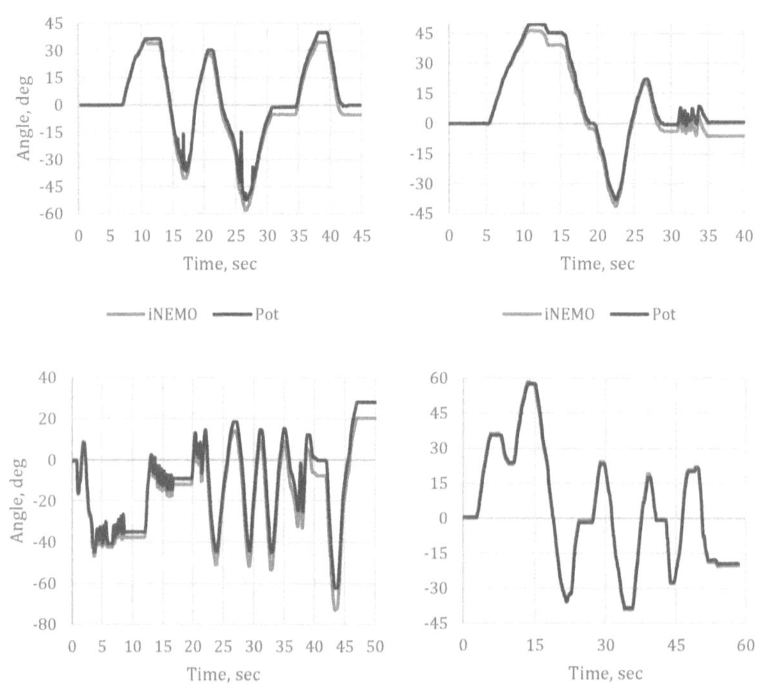

Figure 31: iNEMO's yaw estimate vs the potentiometer's yaw for various rotation sequences

It can be observed that the yaw estimates were initially relatively accurate until the changes in angular position became abrupt. After a relatively high frequency angular motion was applied, the yaw estimate's margin to the actual value grew, deviating from the potentiometer's measurements by about 10 ° by the end of the sampling period (as seen in the both top graphs and the bottom left graph of Figure 31). The bottom right graph however showed that the estimate's drift in response to a low frequency change was significantly lower than the responses to high frequency changes.

To better the yaw's zero rate performance, the drift was compensated for subsequently. This was achieved by removing drift during every iteration of the filter, i.e. by subtracting the following from the yaw rate's integral:

$$16.2\,°/hr \times sampling\ time\ step\ (hr) \hspace{3cm} \{2.21\}$$

The following is a graph of the estimate after implementing the drift compensation on the iNEMO:

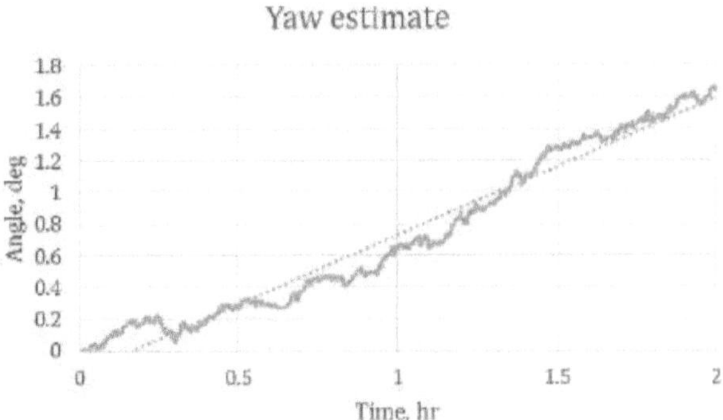

Figure 32: New zero rate drift over a 2-hour period

As shown in the above figure, the maximum zero rate drift was found to be about 1.7 °, and after linearizing the curve, a zero drift rate of about 0.87 °/hr was obtained. This is a significant drop-off from the previous rate of 16.2 °/hr. The following were the estimates obtained from the new implementation.

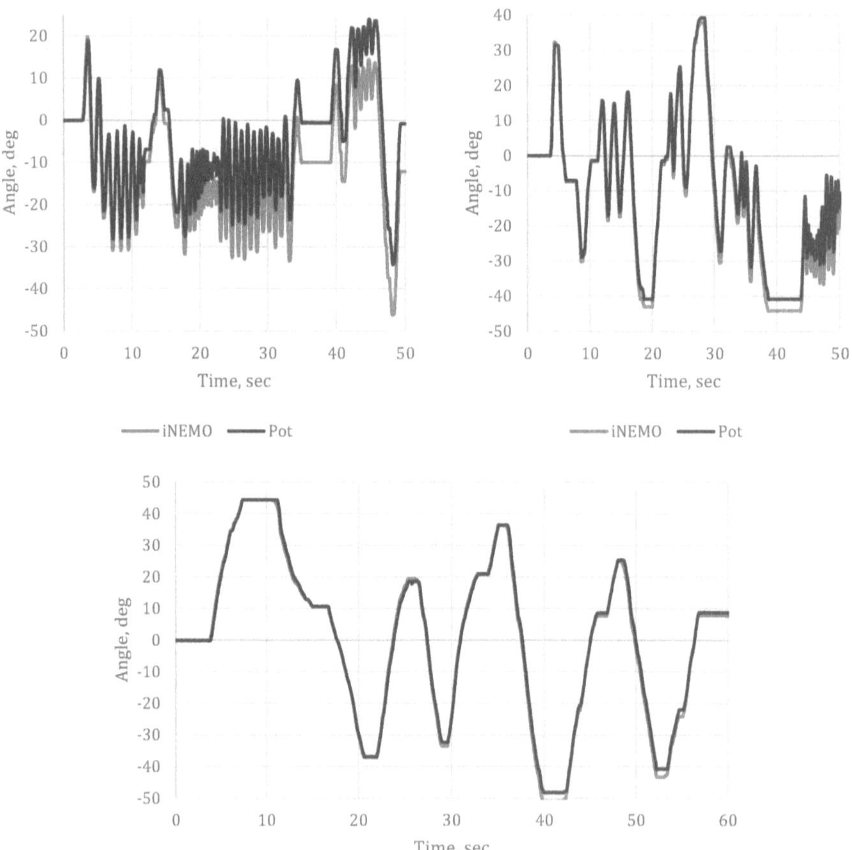

Figure 33: Yaw estimates after compensation for zero rate drift

As expected, the new estimates behaved in the same manner as the previous estimates: less accurate when motion is of the high frequency type. This loss of accuracy was attributed to the filter's sampling frequency as yaw was based solely on angular rate integration. At shorter sampling periods, the estimates were expected to be more accurate. On the other hand, the drift compensation would ultimately be beneficial in the long term to the yaw estimation in form of minimum drift.

4.3.2 Roll and Pitch Estimation

Roll and pitch angles were estimated by setting the gimbal to a specific angle and observing the iNEMO's estimate over a period of time. For example, the pitch gimbal with the iNEMO was set to 30 ° while roll was set to 0 ° for about 50 seconds. This procedure was also followed again but with pitch at about 0 ° while roll was at 10 °. The following table shows the deviation of each tilt angle estimate over the test period. This was carried out to test the accuracy of the filter's tilt angle estimates.

Table 17: Standard deviation for each tilt angle estimate

	Roll	Pitch
Test 1	0.0917 °	0.0932 °
Test 2	0.0891 °	0.0866 °
Average	0.0904 °	0.0899 °

From the standard deviation table, both roll and pitch estimates resulted in average accuracies well within the typical tilt angle accuracy of 1 °. More tilt angle tests were carried out to observe the performance of the filter. Both roll and pitch were separately estimated by rotating the iNEMO randomly about an axis. The rotation was performed with a set square to ensure the estimates settle at the set square angles: 60 ° and 30 °. The following are the result from these rotations.

Figure 34: Estimating roll and pitch through some rotation sequences using a 90-60-30 set square

The result for iNEMO's roll estimate is as follows:

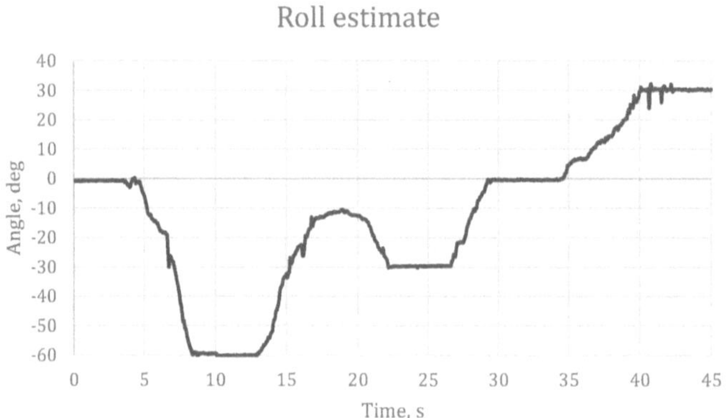

Figure 35: iNEMO's roll output through the rotation sequence

The result for iNEMO's pitch estimate is as follows.

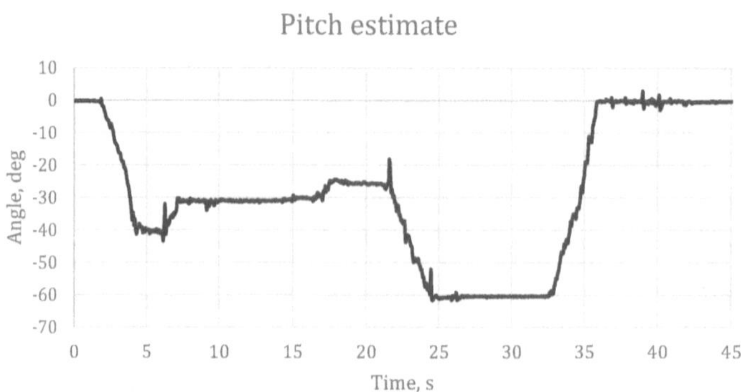

Figure 36: iNEMO's pitch output through the rotation sequence

The tilt angle estimates were quite sensitive to vibrations. This was attributed to the fact that the tilt angle estimates were partly based on measurements from the accelerometer which is capable of detecting really small amplitude accelerations as small as 0.001 g.

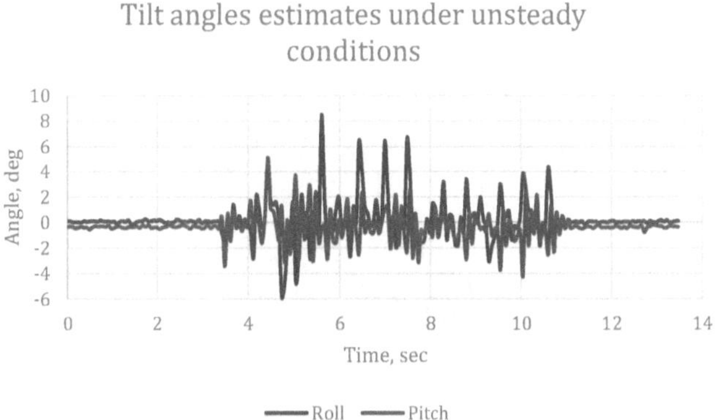

Figure 37: Tilt angle estimates when iNEMO was subjected to small-amplitude vibrations

This sensitivity can be observed from the Figure 37 where a disturbance was introduced by randomly tapping the table on which the iNEMO was placed. The roll and pitch angles were wildly off but settled back to their current values when steady state was restored. To mitigate for unsteady forces introduced into the system, the magnitude of the accelerometer's measurement was employed. If only gravity is present, i.e. system is in steady state, the measured acceleration magnitude would ideally be 1 g. Assuming an allowable deviation of about ±0.05 g from the ideal magnitude, a threshold of 1 ± 0.05 g was therefore implemented such that if a measured acceleration magnitude was outside this range, an unsteady state was assumed and the tilt angle estimates were based entirely on the gyroscope's measurements. After implementing this on the iNEMO and subjecting it to a similar condition under which the Figure 37 estimates were obtained, the following roll and pitch outputs were provided by the iNEMO.

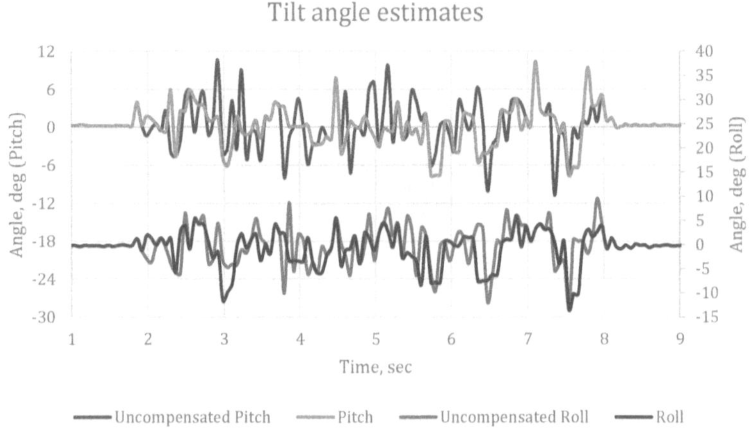

Figure 38: Tilt angle estimates after acceleration threshold implementation

The figure shows that the acceleration cut-off rejects some of the vibration effects on the tilt angle estimates, though not entirely effectively. Vibrations picked up as rotation by the gyroscope introduced some error into the rate integration which ultimately resulted in inaccurate tilt angles.

4.3.3 Initial Angular Position Setting

The Seahog's pilot is given the option to set the initial yaw position based on the reference coordinate system in use. The initial roll and pitch angles were assumed to be estimated accurately by the filter but the yaw always started at zero because its estimate was entirely based on angular rate integration. The initial yaw angle could be an angle relative to the earth's true north sourced from a gyrocompass. Initial Euler angle characterization was achieved on the iNEMO by using the estimated tilt angles and the specified initial yaw angle to compute an equivalent quaternion within the first ten seconds of the iNEMO being online. The initial angles stated in the following graphs are in the order (ψ, θ, ϕ).

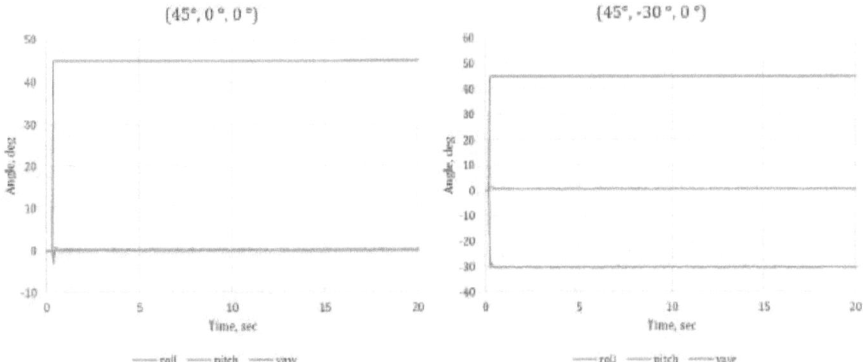

Figure 39: Specifying non-zero initial conditions of the Euler angles

The iNEMO was held in a stationary position which is why the Euler angles in the graphs above were unchanged after ten seconds. This has shown that setting an initial Euler position is a straightforward task. The potentially challenging factor would be the integration of the external device or system upon which the initial yaw conditions will be based with the iNEMO.

4.4 Conclusion and Recommendations

After testing several AHRS algorithms under different conditions, an EKF algorithm capable of providing roll and pitch estimates with accuracies that could match commercial offerings was selected and successfully tested and implanted on the iNEMO. While unsteady forces act on the Seahog or during the transient state, a condition based on the measured Seahog's acceleration was implemented to ensure that the tilt angles were based on the angular rate only. This would ensure insensitivity to the unsteady forces and therefore offer a consistent stream of steady tilt angle estimates. Considering the slow moving nature of the Seahog, its performance was deemed sufficient.

A yaw angle estimation algorithm (which was eventually based solely on angular rate integration) with a minimal drift of about 0.87 °/hr was implemented on the iNEMO. This represented a growth significantly less than 1% of the yaw's full range per hour. The estimate's accuracy is however dependent on the frequency of change in angular position. The faster the change, the less accurate the yaw estimate. This was sufficient for preliminary implementation and use on the Seahog as it is a slow moving vehicle. However, if the opportunity arises however for additional design work on the Seahog and hardware procurement, an acoustic system or a gyrocompass would be alternative yaw estimating devices. The gyrocompass would be preferable as it is a more compact and small unit, it directly outputs heading angles relative to the true north and it would not intrude on marine life like the acoustic-based system would. Unlike the magnetometer, it is non-magnetic therefore insusceptible to magnetic distortions. It also locates earth's true north rather than the magnetic north located by the magnetometer.

As stated in section 4.2.10, under scenarios 1 and 5 which were the sinusoidal and non-zero linear acceleration test conditions respectively, some of the Bayesian filter responses produced errors which were sinusoidal in nature. An instance of this is shown in Figure 40.

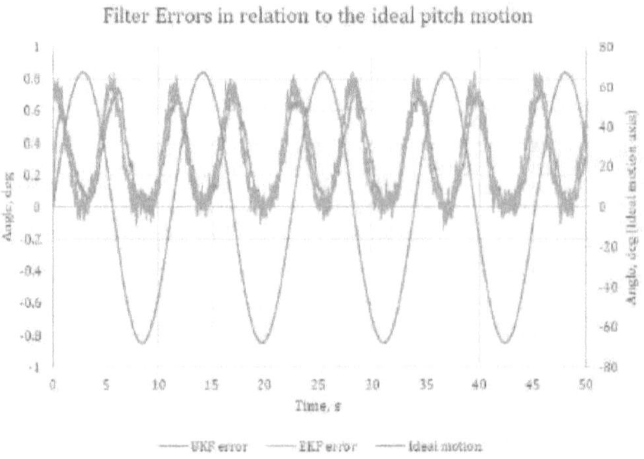

Figure 40: An extract from scenario 1 displaying the EKF and UKF's sinusoidal errors

This behaviour could be investigated further. Not all the Bayesian filters exhibited this characteristic in the test scenario, for example, the SS-UKF under scenario 5 seemed to produce random errors unlike the EKF and UKF. The test scenarios under which this behaviour occurred were dissimilar, i.e. one was a sinusoidal while the other was a step motion. Scenario 2 was also based on sinusoidal motion but none of the Bayesian filters display this characteristic. In this scenario however, the CF error profile does appear to be a wave form. This might be a transient response that would dissipate at a time beyond the simulation period unlike the Bayesian filters' error profile under scenarios 1 and 5 in which the wave forms appeared to have constant amplitudes. Different motion amplitudes and frequencies could be implemented to observe and analyse the Bayesian filters' responses. Different sample frequencies could also be implemented on the filters to potentially quantify the filter output relationship with the sampling frequency.

5 A LITERATURE STUDY ON ADDED MASS

5.1 Hydrodynamic Effects on an Underwater Vehicle

Hydrodynamic forces and moments are the dynamic effects on underwater or surface vehicles, caused by the surrounding body of fluid. In this study, hydrodynamic effects on underwater vehicles were of much relevance, therefore most of the subsequent work shown in this chapter focuses on content pertaining to underwater vehicles. There are hydrodynamic effects on underwater vehicles when they surface (before submersion or after re-surfacing), but these effects were not deemed important enough at this stage of development as the Seahog's underwater operational dynamics were seen to be more significant for the manoeuvring and control of the Seahog.

The control of an underwater Remotely-Operated Vehicle (ROV) can be significantly compromised if the effects of the dynamics between the ROV and the fluid it is submerged in are not taken into account. The hydrodynamic coefficients are quantities obtained from forces and moments due to the fluid acting on a partially or fully submerged body. These forces can subsequently be placed under two groups, radiation-induced forces and environmental disturbances such as surface waves and winds [14].

The radiation-induced forces are forces on the body due to the excitation of the surrounding fluid. They are composed of the following components:

- The added mass (also known as induced mass) effect and the Basset-Boussinesq force due to relative motion of the body in a fluid body.

- Potential damping due to the transfer of energy by surrounding fluid and radiated waves.

- Restoring forces acting on the submerged object due to its weight and buoyancy.

For rigid bodies, using the Newton-Euler formulation, the effect of these forces and moments can be mathematically represented as a superposition of components of forces and moments [14]:

$$\tau_R = \underbrace{-M_A\dot{v} - C_A(v)v}_{\substack{added\ mass \\ effect}} - \underbrace{D(v)v}_{\substack{total\ damping \\ effect}} - \underbrace{g(\eta)}_{\substack{restoring \\ forces}} \qquad \{5.1\}$$

where τ_R are the radiation-induced forces, $g(\eta)$ is the restoring (righting) force due to buoyancy and weight, M_A is the added mass matrix, $C_A(v)$ is the Coriolis matrix and $D(v)$ is the total superimposed damping effect including radiation-induced potential damping, skin friction, wave drift damping and damping due to vortex shedding [14]. $C_A(v)$ and $D(v)$ are both a function of the 6 degrees of freedom (DOF) velocity vector v. The above equation can be shown to fit with the 6 DOF equation of motion of the vehicle as follows [14]:

$$(M_{RB} + M_A)\dot{v} + (C_{RB}(v) + C_A(v))v + D(v)v + g(\eta) - \tau_E = \tau \qquad \{5.2\}$$

where M_{RB} and C_{RB} represent the 6 × 6 rigid body inertia, and Coriolis and centripetal matrices respectively, τ_E is the force due to environmental factors and τ is the propulsion force and moment matrix of the vehicle [14]. Note that the aforementioned force and moment equations are general and apply to both underwater and surface vehicles. A detailed formulation for the Coriolis and centripetal force matrix can be found in the book by Fossen [14].

5.1.1 Added Mass and Moment of Inertia

Added mass forces and moments are the added "*induced mass and moment*" effects of the surrounding fluid on an accelerating body. This phenomenon is common to both partially and fully submerged vehicles. This effect is propagated when a body's motion causes the surrounding body of fluid to oscillate with different fluid amplitudes all in phase with the forced harmonic motion of the vehicle [14]. The fluid's oscillation amplitude does decay the farther it is from the moving body, causing it to become negligible [14]. Added mass is almost always positive for rigid fully submerged bodies, i.e. bodies in ideal fluid regions (usually away from the free surface) with no incident waves, fluid currents and also independent of wave circular frequencies [14].

A Lagrangian formulation of the added mass is described by Fossen [14]. Added inertia quantities for fully submerged vehicles are found to be always positive, i.e. $M_A > 0$, but for bodies close to the free surface or exposed to certain wave circular frequencies, negative added mass values have been recorded [14].

5.1.2 Hydrodynamic Damping

Damping dissipates kinetic energy from an accelerating body. The hydrodynamic damping forces on a submerged vehicle can be as a result of one or a combination of the phenomena defined in the following table. The type (s) of hydrodynamic damping force on a submerged vehicle depends on a number of factors such as depth travelled, proximity to the seabed or free surface, frequency of oscillation, fluid wave or current propagation, geometry of body and surface roughness [52]. Drag experienced by a fully or partially submerged vehicle can be a combination of the damping types on Table 18 depending on the aforementioned factors affecting damping.

Table 18: Hydrodynamic damping types

Damping type	Definition
Radiation-induced damping	Also known as potential damping is as a result of the vehicle being forced to oscillate at the wave's excitation frequency [14]. Its effect is significant on bodies operating at the fluid body's surface and negligible on vehicles submerged to great depths compared with other damping effects [14]. This dissipative effect is frequency dependent.
Skin friction	This damping effect is related to boundary layer formation. It can be represented linearly (linear skin friction) on slow moving vehicles due to laminar boundary layer theory and

	non-linearly (quadratic skin friction) on high frequency vehicles due to turbulent boundary layer theory [14].
Wave drift damping	Fossen [14] defines wave drift damping as the added resistance for surface vessels advancing in waves. This is derived from the 2nd order wave theory [14].
Damping due to vortex shedding	Frictional forces act on a body moving in a viscous fluid. The drag force acting on a body due to vortex shedding can be expressed as [14]: $$D(u) = -\frac{1}{2}C_D(R_n)A|u|u$$ Where C_d is the drag coefficient relevant to a certain projected cross-sectional area A of a body moving at a velocity u. C_d is a function of Reynolds number R_n which is an indicator of whether a fluid flow relative to a body is laminar or turbulent.

As stated earlier, when a body accelerates or decelerates in a fluid, this results in the impacting of kinetic energy on the surrounding fluid. The unsteady forces on the body due to the fluid's kinetic energy can also be divided into two groups, namely: the added mass effect and the Basset-Boussinesq force.

The Basset-Boussinesq force is generally neglected as it is difficult to implement practically [53]. It is the force due to the lagging boundary layer development with the acceleration of a body moving through a fluid. It is also referred to as the history force as it is a force associated with the change in the fluid's flow pattern over time. The force can be significant when the body is accelerated at a high rate relative to the fluid [53]. Due to the above-mentioned points, the Basset-Boussinesq force is going to be taken as negligible throughout the subsequent analysis as the Seahog is not a fast moving vehicle.

The hydrodynamic effects can be summed up mathematically as:

$$\tau_R = -M_A\dot{v} - C_A(v)v - D_P(v)v - g(v) + g_0 \qquad \{5.3\}$$

5.1.3 Methods of Determining Hydrodynamic Forces

The hydrodynamic coefficients are the hydrodynamic force and moment variables, including the added inertia coefficients and damping coefficients obtained experimentally or through some theoretical estimation. In the real world, these values may vary depending on conditions such as fluid properties like flow rate, temperature, density, circular wave frequency and vehicle properties like depth travelled, velocity, geometry, surface roughness [52]. However, the coefficients are generally taken to be constant for underwater vehicles due to the fixed or slow-changing nature of some of the aforementioned characteristics. When the vehicle is performing a manoeuvre involving acceleration relative to the fluid, that is, the vehicle is in an unsteady state, the effects of the acceleration on the surrounding fluid is a lot more significant than the fluid's circular motion along the vehicle's surface. For this reason, the use of fluid flow models ignoring circular wave effects when estimating hydrodynamic forces and moments, especially added inertia, is generally accepted.

There are multiple methods of obtaining hydrodynamic masses and moment of inertia, including both experimental and theoretical approaches. The theoretical methods, either analytic or numeric in nature, are generally computational because of the amount of work and time required to solve the equations involved, especially when dealing with complex geometries. For basic geometries like spheres, cuboids, parallelepiped etc., analytical formulae such as the formulae found in the technical documentation by DNVGL [52] for obtaining added inertia and damping forces are available. A sample of these solutions is shown in Figure 42.

The theoretical basis for solving for hydrodynamic forces is on the fluid flow being *inviscid*, which therefore validates the use of *potential flow theory*. A fluid that is inviscid had no viscosity (friction) and ensures that the conservation of fluid momentum is satisfied. Potential flow describes the fluid velocity field around a vehicle as a gradient of a scalar velocity potential function [54]. When dealing with added mass, the fluid modelled is generally an *ideal* fluid described using the potential flow theory. An ideal fluid is an irrotational and incompressible fluid. An irrotational flow describes a linear fluid flow field without rotation. Therefore, the most ideal fluid model would be incompressible, frictionless and without vortices. This might seem like an over simplification of the vehicle-fluid interaction but it is widely seen as accurate enough to be used in preliminary academic or industrial simulation or added mass estimation. Unlike the potential flow which is conservative, the Navier-Stokes equations are dissipative and can be used to represent a viscous flow. Linearity however does not hold for fluid flow represented using Navier-Stokes equations [55]. The phenomenon of the ideal fluid flow where the drag force experienced by a body moving at constant velocity is zero is referred to as D'Alembert's paradox [56]. This is contradictory to the normal body-fluid interaction: resistance is experienced by a body in motion due to the surrounding fluid.

Reynolds number plays a big part in determining which flow model to apply for flow linearity to hold. Reynolds number can be defined as a dimensionless quantity representing a ratio of inertial forces to viscous forces. Flow linearity ensures superposable forces from acceleration effects in different directions. For flows having low Reynolds numbers i.e. low flow velocities, of Reynolds number $Re \ll 1$, a *Stokes flow* model is required, while for high Reynolds number, a potential flow model would be necessary [55]. Fluids such as slurries would be modelled by the Stokes flow due to their high viscosity level. The case relevant to the Seahog is the high Reynolds number case.

The following formulation aims to provide an overview of the derivation of the added mass equation based on potential flow theory. The fluid is taken to be initially at rest before being excited into motion. The fluid's linear momentum $\vec{L}$ at an arbitrary point can be represented by the following triple integral, where ρ represents density, u is a unit velocity, V is the fluid's volume and Φ is the velocity potential function corresponding to the unit velocity. The velocity potential function Φ is assumed to satisfy the basic fluid mechanics laws of mass and momentum conservation. The potential Φ therefore also satisfies the Laplace's equation $\nabla^2\Phi = 0$ since it is an incompressible fluid.

$$\vec{L} = \iiint_V \rho\vec{u}\, dV = \iiint_V \rho\nabla u\Phi\, dV \qquad \{5.4\}$$

Focussing on obtaining added mass along 1 axis (x-axis in this instance), the momentum equation can be represented using Green's theorem as,

$$L_x = \iint_S \rho u \Phi n_x \, dS \qquad \{5.5\}$$

where dS represents the infinitesimal surface and n_x is the unit normal vector. The force on the fluid along the x-axis due to an accelerating body is $F_x = m_x \dot{u}$. Force and momentum can be related through the equation,

$$\Delta L_x = \int_T F_x \, dt \qquad \{5.6\}$$

Applying the kinematic boundary condition to the fluid yields the following expression,

$$\frac{\delta \Phi}{\delta n} = n_x \qquad \{5.7\}$$

Therefore, the equation for the added mass along the x-axis can be obtained from a combination of equations 5.6 and 5.7,

$$m_x = \rho \iint_S \Phi \frac{\delta \Phi}{\delta n} \, dS \qquad \{5.8\}$$

This equation can then be generalized to include all 6 DOF, where i, j represents the components of an added mass tensor,

$$m_{ij} = \rho \iint_S \Phi_j \frac{\delta \Phi_i}{\delta n} \, dS \qquad \{5.9\}$$

Birkhoff's book on hydrodynamics [57] (pp. 148-160) provides a more detailed derivation of the added mass equation.

There are various approaches available in determining or estimating a marine vehicle's hydrodynamic added forces. Analytical added mass solutions for basic geometries are readily available. Brennen [55], Chung and Chen [56] and DNV [52] have documented analytical added mass and inertia solutions for some basic 2D and 3D geometries. The solutions in these documents are valid for fully submerged bodies that are operating in a stationary and ideal fluid of infinite volume, i.e. the body is far away from the fluid's boundaries. The analytical solutions provided are only for the diagonal elements of the added mass matrix M_A (see previous section on added mass for more details on this matrix). The matrix is a 6×6 matrix but because of the system being superposable i.e. total added mass along an axis is a summation of added masses due to unsteady motion along all 6 axes, and conservative [55], it consists of 21 distinct elements. The other missing elements can be obtained by applying the theorem of reciprocity to M_A, i.e. $M_A = M_A^T$. Geometric symmetry can further reduce the number of distinct added mass matrix elements. For example, for the body in Figure 41 having two planes of symmetry, seven distinct added inertia elements exist.

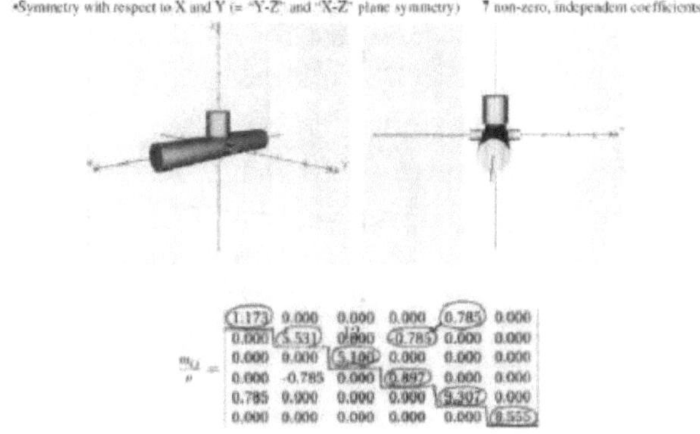

Figure 41: Added mass coefficient matrix [58]

Another analytical approach to estimating added mass is by applying the *strip theory*, also referred to as the *slender body approximation*. This method involves slicing a submerged 3D shape into thin cross-sectional layers whose added masses could be individually estimated. These values are then summed up i.e. integrated over the entire length of the 3D shape, to provide an estimate for the 3D shape. It can be applied to fully or partially submerged bodies. This solution only works for slender bodies i.e. bodies having significantly longer longitudinal lengths than their diametric length. The following equation, where A_{ij} is the 3D added mass in the i direction caused by acceleration in the j direction and a_k is the 2D added mass of cross-sectional slice k, over the entire body length L.

$$A_{ij} = \int_L a_k \, dx \qquad \{5.10\}$$

It is obvious that the slender body theory is simple when the body has a simple and uniform cross section. It becomes more challenging for a complex body. The strip theory however, because it relies on the added inertia of 2D shapes, cannot provide added inertia estimates along the body's longitudinal direction, as 2D shapes generally do not have added inertia values along that axis. The Society of Naval Architects and Marine Engineers provide a more thorough examination of the strip theory [59] (pp. 53-73).

Shape of Cross Section	Direction of Motion†	Hydrodynamic Mass (per unit length)	Hydrodynamic Mass Coeff.†† (c_n or c_t)
Solid Body in an Infinite Fluid — Circle (2R)	Any Direction	$\rho\pi R^2$	1.0
	Rotational	0	0
Ellipse (2a, 2b)	Vertical	$\pi\rho b^2$	$\frac{b}{a}$
	Horizontal	$\pi\rho a^2$	$\frac{a}{b}$ (See Fig. 11)
	Rotational	$\frac{\pi}{8}\rho(b^2-a^2)^2$	$\frac{[1-(a/b)^2]^2}{2(a/b)[1+(a/b)^2]}$ (See Fig. 14)
Semi-circle (R)	Rotational	$\left(\frac{4}{\pi}-\frac{1}{4}\right)\rho\pi R^4$	$\frac{16}{\pi^2}-1$
Thin-strip (h, 2a) (h << a)	Vertical	$\pi\rho a^2$	$\frac{b}{2}\left(\frac{a}{h}\right)$
	Rotational	$\frac{\pi}{8}\rho a^4$	$\frac{3}{16}\pi\left(\frac{a}{h}\right)$

Figure 42: Excerpt from the technical report by Chung and Chen [56] of some analytical solutions for some basic 2D shapes

The following is a full matrix of distinct added mass/moment of inertia components derived from the strip theory. The rest of the matrix can be completed using symmetry.

$$A_{22} = \int_L a_{22}\, dx \quad A_{23} = -\int_L a_{23}\, dx \quad A_{24} = \int_L a_{24}\, dx \qquad\qquad A_{26} = \int_L x a_{22}\, dx$$

$$A_{33} = \int_L a_{33}\, dx \qquad\qquad A_{35} = -\int_L x a_{33}\, dx$$

$$A_{44} = \int_L a_{44}\, dx \qquad\qquad A_{46} = \int_L x a_{24}\, dx$$

$$A_{55} = \int_L x^2 a_{33}\, dx$$

$$A_{66} = \int_L x^2 a_{22}\, dx \Big]$$

Various Computational Fluid Dynamics (CFD) programs exist in the marketplace that have been used to model and perform hydrodynamic force and drag analyses of a vehicle-fluid interaction. They generally employ a Finite Element Method (FEM) to model the structure and fluid flow. FEM is a numerical method applied in continuum mechanics that provides an approximate solution to a differential equation which does not have an analytical solution, or whose solution is difficult to obtain by traditional methods [60]. This is a fine approach for performing structural and drag analysis but the added mass is generally calculated using another method. Traditionally, the added mass would be obtained analytically, independent of the structural model [61]. Alternatively, a numerical method called the Boundary Element Method (BEM) is employed. BEMs differ to FEMs by the fact that boundary domains are discretized rather than entire physical domain discretization otherwise performed in an FEM analysis. BEM is preferable for solving for added forces because they can handle complex geometries, provide more accurate results and can work with domains extending to infinity [62]. Two of the standard BEM methods are discussed below based on the computational work performed by Menk et al. [63] on a fluid-ship hull interaction:

- *Full hydrodynamic mass matrix method* – This involves coupling an already existing body's Finite Element (FE) model with the FE of the surrounding fluid. The overall Laplace equation is then solved using coupled boundary conditions. This approach however leads to the cubic scaling of the required computational time and a quadratic scaling of memory usage with the additional fluid FE panels.

- *Lewis method* – This approach approximates the added fluid inertia by analysing the 2D flow across the body's cross-section. This method assumes *slender body* restrictions hold. A drawback of this approach is that it does not provide the added force distribution along the body as only the total force on each cross-section can be computed.

These methods are computationally expensive (with regards to memory requirements and runtime), especially when dealing with really complex and large systems. More computationally efficient approaches exist for determining added forces such as the *projection approach, multi-pole method, panel method* etc. [63]. These methods can be collectively referred to as the Fast Boundary Element Method. A more efficient method not only means less expensive computational power requirement, it also means a more accurate analysis can be performed, as more boundary elements can be used leading to approximations closer to the exact solution.

A popular commercial fluid-body interaction analysis software used for estimating added mass is called WAMIT (Wave Analysis MIT). WAMIT is a linear analysis radiation/diffraction program evaluating the interaction of surface waves with different types of floating or submerged bodies [64]. WAMIT is a panel based program and was originally developed by researchers at Massachusetts Institute of Technology. It cannot be used to actually model the body/vehicle requiring analysis, so the part has to be modelled externally and imported into WAMIT. A common 3D modeller used is MultiSurf.

The methods previously discussed apply to bodies far from the fluid's boundary layers. There are ways to determine added mass of bodies close to the free surface or other boundary layers. Zhou et al. [65] performed numerical analysis based simulations on cylinders of semi-circular, rectangular and triangular cross-sections to observe the effects on sway, heave and roll added mass of changing proximity to seabed and narrowness of the fluid body. It was found that the added mass generally increased with closeness to a boundary. A complex potential flow was employed in their numerical analysis. For their analysis, *wide water* was defined as ratios in excess of water height to diametric radius ratio of $H/r_0 = 10$ and water width to diametric radius ratio of $b/r_0 = 10$. Clarke [66] employed the method of conformal mapping to determine the added mass on a ship hull in shallow water. Conformal mapping involves mirroring of the relevant domain about the water surface line, leading to a double body. The hull was approximated as a semi-circular cylinder. This approach only works in the case of no surface wave frequencies.

5.2 The Seahog's Current Hydrodynamic Model Review

Finbow developed the Seahog's simulation employing the coefficients he obtained [2]. For this reason, a review of his work, particularly on the added mass determination was necessary.

5.2.1 Added Mass and Moment of Inertia

The current added mass values in the Seahog's simulation were obtained through theoretical estimation. The hydrodynamic effects on the Seahog's top half were taken as the most significant contributor of the added mass. Consequently, this led to the Seahog's overall geometry being simplified into a cuboid, neglecting the lower frame and its horizontal thrusters. This has obvious limitations as the contributions of the lower frame and horizontal thrusters of the Seahog are neglected, meaning that the current added mass figures in the surge and sway direction are expected to be under-estimates, while the added mass estimate in the heave direction is expected to be an over-estimate due to the model used having more surface area perpendicular to the heave motion direction than the Seahog. An assumption that the Seahog is moving in a still fluid body, therefore causing the fluid around it to accelerate is an important assumption as the formulae used are valid within this scenario.

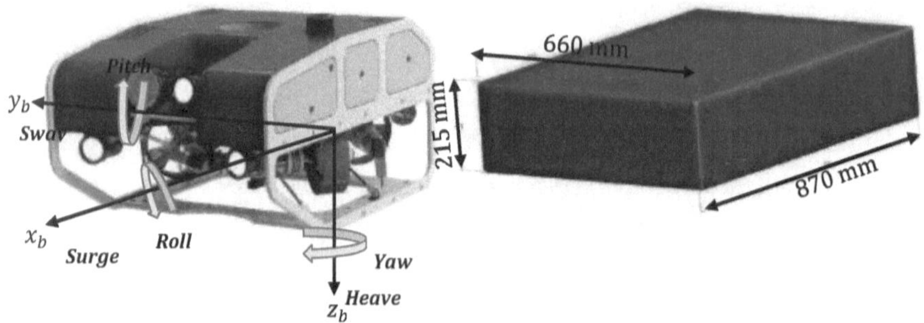

Figure 43: Left - Seahog, Right - Simplified 660 mm x 870 mm x 215 mm (L x B x H) version used in the added mass estimation

Two methods were employed in obtaining the translational added mass terms: through a 2D analysis and a 3D analysis. The 2D added masses were obtained using the type 1 formula in Figure 44. For the 3D analysis, types 2 and 3 from Figure 44 were used to estimate the heave added mass (taking the average of both) and type 3 was used for the surge and sway added mass estimation. The average of both methods were then taken as the added masses in each respective linear direction.

Type of Analysis	Direction of Motion	a/b	k	Reference Area/Volume
1 Rectangular plates	Vertical	0.1	2.23	$A_R = \pi a^2$
		0.2	1.98	
		0.5	1.70	
		1	1.51	
		2	1.36	
		5	1.21	
2	Vertical	0.40	0.801	$V_R = \dfrac{\pi}{4}a^2b$
		0.50	0.757	
		0.63	0.704	
		0.67	0.690	
		0.80	0.642	
		1.00	0.579	
3	Vertical	0.17	0.13	$V_R = a^2b$
		0.20	0.15	
		0.25	0.19	
		0.33	0.24	
		0.50	0.36	
		1.00	0.68	

Figure 44: Added mass coefficient data used in analysis [2]

The translational added masses were calculated using the following formulae, where ρ is the fluid density and L is the longitudinal length of the cuboid. k is the coefficient corresponding to length ratio a/b as shown in Figure 44.

$$m_A = \rho k A_R L \ \text{(2D analysis)} \tag{5.11}$$

$$m_A = \rho k V_R \ \text{(3D analysis)} \tag{5.12}$$

The rotational added moment of inertia were obtained through the strip theory. As stated in a previous section, the definition of a slender body is a body possessing a characteristic longitudinal length significantly longer than its cross-sectional diametric length, i.e.

$$longitudinal\ length \gg diametric\ length$$

The formula relating the added moment of inertia about the x-axis $K_{\dot{p}}$ to known 2D translational added masses is:

$$K_{\dot{p}} = \frac{1}{12}\left(Z_{\dot{w}}^{2D} B^3 + Y_{\dot{v}}^{2D} H^3\right) \tag{5.13}$$

where $Z_{\dot{w}}^{2D}$ and $Y_{\dot{v}}^{2D}$ are the 2D cross-sectional added masses (unit kg/m) along the z and y axis respectively, B is the breadth, H is the height and L is the length of the cuboid. Since the geometry is uniform, i.e. cross-section is the same along the entire length, the total inertia $K_{\dot{p}}$ was obtained by multiplying by the longitudinal length. This same format was followed when estimating the added moment of inertia about the y and z axes, $M_{\dot{q}}$ and $N_{\dot{r}}$ respectively:

$$M_{\dot{q}} = \frac{1}{12}\left(Z_{\dot{w}}^{2D} L^3 + X_{\dot{u}}^{2D} H^3\right) \tag{5.14a}$$

$$N_{\dot{r}} = \frac{1}{12}\left(X_{\dot{u}}^{2D} B^3 + Y_{\dot{v}}^{2D} L^3\right) \tag{5.14b}$$

The added mass matrix M_A therefore consisted of 6 distinct added mass and moment of inertia in the surge, sway, heave, roll, pitch and yaw directions; all on the diagonal of the matrix.

$$M_A = \begin{bmatrix} X_{\dot{u}} & 0 & 0 & 0 & 0 & 0 \\ 0 & Y_{\dot{v}} & 0 & 0 & 0 & 0 \\ 0 & 0 & Z_{\dot{w}} & 0 & 0 & 0 \\ 0 & 0 & 0 & K_{\dot{p}} & 0 & 0 \\ 0 & 0 & 0 & 0 & M_{\dot{q}} & 0 \\ 0 & 0 & 0 & 0 & 0 & N_{\dot{r}} \end{bmatrix} = \begin{bmatrix} X_{\dot{u}} & Y_{\dot{v}} & Z_{\dot{w}} & K_{\dot{p}} & M_{\dot{q}} & N_{\dot{r}} \end{bmatrix}^{D} \qquad \{5.15\}$$

The added mass matrix M_A is valid as making a three planes of symmetry assumption for a low speed vehicle leads to negligible cross-coupling effects, i.e. the off-diagonal effects of the matrix are negligible.

Table 19: Added inertia values

Term	Added Inertia Value
$X_{\dot{u}}$	42.1 kg
$Y_{\dot{v}}$	57.2 kg
$Z_{\dot{w}}$	359.9 kg
$K_{\dot{p}}$	7.0 kgm²
$M_{\dot{q}}$	16.0 kgm²
$N_{\dot{r}}$	4.1 kgm²

For a more detailed derivation of these figures, see chapter 5 of Finbow [2].

5.2.2 Hydrodynamic Damping

The hydrodynamic damping forces on the Seahog were obtained by performing computational fluid analysis on SolidWorks. SolidWorks achieves this by solving time-dependent Reynold-averaged 3D Navier-Stokes equations which describes turbulent flows, by employing the Finite Volume Method [67]. It was deemed necessary to perform the SolidWorks drag simulations over the velocity range of the Seahog in all decoupled 6 degrees of freedom (DOF), therefore the maximum speeds in each direction were determined. This was achieved by using drag simulations, where the known maximum thrust in a single DOF of interest was opposed by a drag force of the same magnitude. The following table details the estimated maximum speed along each DOF.

Table 20: Maximum speed in each DOF

DOF	Direction	Max. thrust	Max. speed
Surge	x	176.8 N	1.5 m/s
Sway	y	162.0 N	0.8 m/s
Heave	z	60.0 N	0.4 m/s
Roll	Φ	16.2 Nm	1.4 rad/s
Pitch	θ	17.7 Nm	1.0 rad/s
Yaw	ψ	63.7 Nm	2.6 rad/s

However, the speed range actually used in the drag simulations were taken to be slightly greater than the estimated figures (as seen on Table 21) based on the premise that forces other than drag and thrust would act on the Seahog. The adjustment made to the relative speeds were not necessary in this instance because while there are other forces acting on the Seahog besides the thrust and skin friction drag, any other radiation-induced effect like the added mass or vortex shedding would only act to reduce the effect of the thrust, thereby lowering the maximum possible speed. This is especially true for added mass as it is always > 0 for completely submerged vehicles [14]. This can be seen in the equations 5.16:

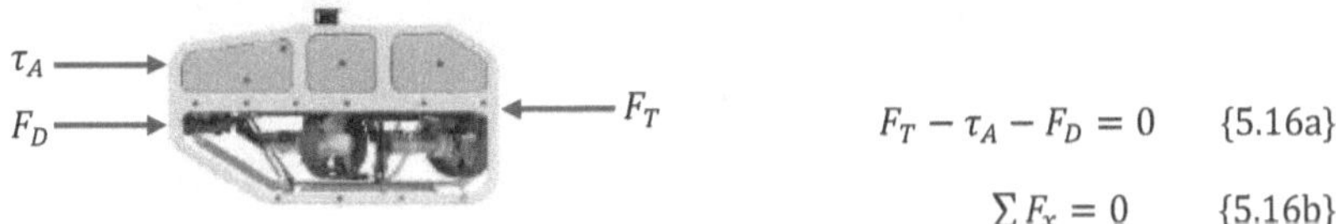

$$F_T - \tau_A - F_D = 0 \qquad \{5.16a\}$$

$$\sum F_x = 0 \qquad \{5.16b\}$$

Figure 45: Horizontal forces on the Seahog

F_T is the total thrust exerted by the thrusters in a single DOF, F_D represents a superposition of the drag forces and τ_A is the sum of the added mass and Coriolis effects.

Table 21: Estimated vs simulation speed comparison

Direction	Adjusted simulation speed range	Max. estimated speed	% difference	Drag force/moment
x	±1.75 m/s	1.5 m/s	16.7%	X (N)
y	±1.00 m/s	0.8 m/s	25%	Y (N)
z	±0.60 m/s	0.4 m/s	50%	Z (N)
ϕ	±1.50 rad/s	1.4 rad/s	7.1%	K (Nm)
θ	±1.20 rad/s	1.0 rad/s	20%	M (Nm)
ψ	±2.75 rad/s	2.6 rad/s	5.8%	N (Nm)

The Seahog model was also modified for the drag simulations. Due to the inherent complexities of the thrusters in modelling drag characteristics, the horizontal thrusters were removed [2]. The propeller of the vertical thruster was also omitted.

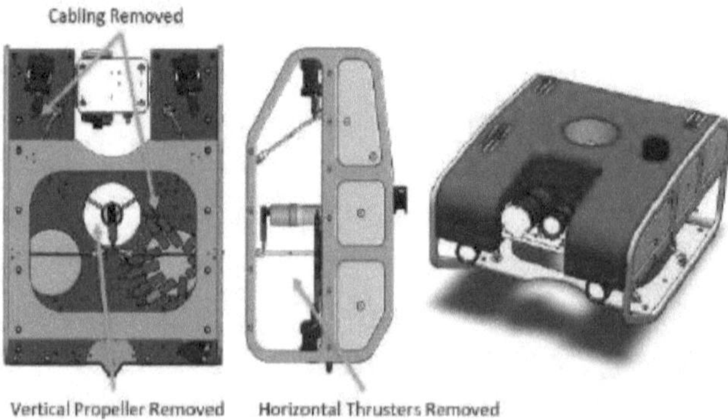

Figure 46: Modified Seahog model for drag simulations [2]

The following table highlights the most important drag simulation parameters as set by Finbow [2]. Notes are given to explain some of the reasoning behind the choices made. These general settings were applied to all drag simulations ran.

Simulation setup parameter	Setting	Note
Surface roughness	0	Unknown surface roughness properties due to incomplete state (cover yet to be designed) of Seahog and different surface materials.
Meshing	8 max	Size factor by which adjacent mesh elements can differ by.
Simulation domain size	10× Seahog overall size	20 times is recommended but 10 was deemed sufficient enough after comparing the results of drag simulations (in the surge, sway and heave directions) between 10× and 20× domain sizes and getting differences less than 6%. Simulation duration time constraints factored in the selection of this setting.
Turbulence intensity	0.1%	Percentage of flow that is unsteady (turbulent). This selection was based on the recommendation of settings well below 1% for slow moving submerged vehicles.
Water density	1000 kg/m^3	Unknown potential testing environment. Chosen because the value was deemed sufficient even though there is a density difference of about 2% between fresh and sea water.

After the setup, drag simulations were performed on the Seahog's SolidWorks model in the direction of each DOF separately. The mesh domain used in each of the linear DOF simulations and the rotational DOF simulations can be seen in the following figures respectively:

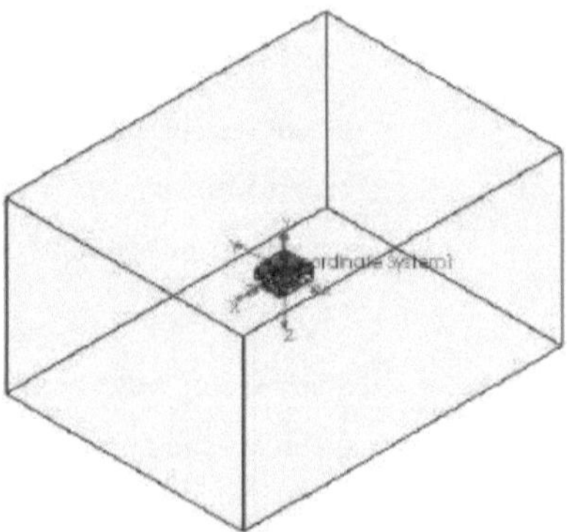

Figure 47: Linear DOF mesh domain [2]

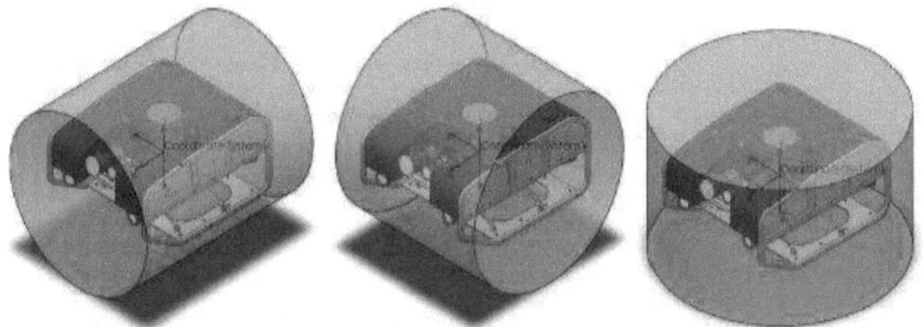

Figure 48: Rotational DOF mesh domains: from left - roll, pitch and yaw respectively [2]

The equations obtained from these drag simulations were then implemented in the Seahog's Simulink simulation. The drag forces and moments in each DOF due to decoupled motion in each DOF were obtained in relation to the relevant velocities (linear or rotational). Therefore, the total drag force or moment along a DOF axis on the Seahog when performing a coupled motion was then implemented as a superposition of all component drag forces/moments due to decoupled motion in the relevant DOF.

5.2.3 Added Mass and Moment of Inertia Review

As stated in section 5.2.1, due to the simplified model used, the estimated added masses and added moment of inertia were expected to vary significantly to their real counterparts. The added masses in the surge and sway directions were expected to be under-estimates due to the omission of surface area and the added mass in the heave direction was expected to be an over-estimate due to the surface area perpendicular to this direction being about 10% more that of the Seahog.

The added moment of inertia were also expected to differ significantly from their actual values due to the aforementioned reasons; they were also derived as functions of the 2D translational added masses through the strip theory. The strip theory is valid for slender bodies; the cuboid does not meet this criteria in some of the cases. For example, the longitudinal length/diametric length ratio is less than one when computing $X_{\dot{u}}^{2D}$.

For these reasons, a different added mass estimation approach that will maintain the geometric properties of the Seahog and produce added mass values close to the true values was to be considered.

6 Experimental Determination of the Added Mass

6.1 Added Mass Estimation Methodology

It was concluded that due to the cost associated with getting a CFD software that can estimate added mass and moment of inertia values, an experimental approach was seen as the most feasible option. The best way to perform hydrodynamic coefficients identification on a marine body would be by analysing data gathered from actual tests performed. One of the most accurate experimental approaches available for identifying all hydrodynamic coefficients involves employing a Planar-Motion Mechanism (PMM) system. PMM is a dynamic system of actuators and sensors capable of monitoring and controlling the velocities and forces on a water vessel. They are typically setup overhead a towing tank (also referred to as a basin). Tests are either performed on the vessels or on small-scaled models, depending on their size, the basin's size and the PMM's capacities. These basins vary in size, depending on the type of testing a research facility/unit undertakes, but in general, they tend to be quite large. For example, the size of the shallow water basin at Ghent University (as seen in Figure 49) is 87.5 m (length) × 7 m (width) × 0.5 m (depth) while a second facility under construction is expected to be 174 m × 20 m × 1 m [68]; MIT has two large basins, 1 small basin and a water tunnel, one of the large basins being of size 30.5 m × 2.4 m × 1.2 m and a 0.5 m square tunnel of length 1.2 m [69].

Another configuration is in the form of a water tunnel with controlled fluid flow. In this case, the vessel or part is held stationary within the tunnel while fluid flows around it. Some of these tunnels have pressure control installed such as the aforementioned MIT water tunnel facility [69]. Though a water vessel's parameter identification can be performed through CFD, when available, basin testing is generally chosen as it is a cheaper alternative when compared to the cost of running a CFD test on such complex structures.

Figure 49: Towing tank at Ghent University, Belgium with small ship model attached to a PMM. Left: [70], right: [68].

Unfortunately, there wasn't access to a basin fitted with the PMM system, so this method was not a feasible approach. It would also be expensive and premature to send the Seahog to a research facility with this system in place at this stage of development. An alternative approach of identifying translational added mass coefficients of an open frame ROV involving free oscillation was proposed by Chin et al. [3]. The following subsection explains this method in detail.

6.1.1 Pendulum Experiment Theory

This approach had been successfully used to estimate the surge, sway, heave and yaw added masses and drag coefficients of a scaled model of an open-frame underwater ROV (similar in size to the Seahog) by Chin et al. [3]. They compared the results of these parameters obtained experimentally with those obtained computationally: WAMIT for the added masses and ANSYS CFX for drag coefficients. This approach has found popularity as an inexpensive and relatively easy to setup method for hydrodynamic coefficients identification.

For the decay test, there are two general configurations possible: the spring-mass setup and the pendulum setup. The pendulum system was preferred to the spring-mass system as the pendulum system would be easier to design to allow strictly one

DOF (the rotation of the pendulum), while the spring-mass would have oscillation components in multiple directions without designing a restraining system to keep the oscillation along one axis.

The body/component being analysed is initially positioned at an angle θ relative to the vertical and released to freely oscillate in the water body initially at rest (as shown in Figure 50).

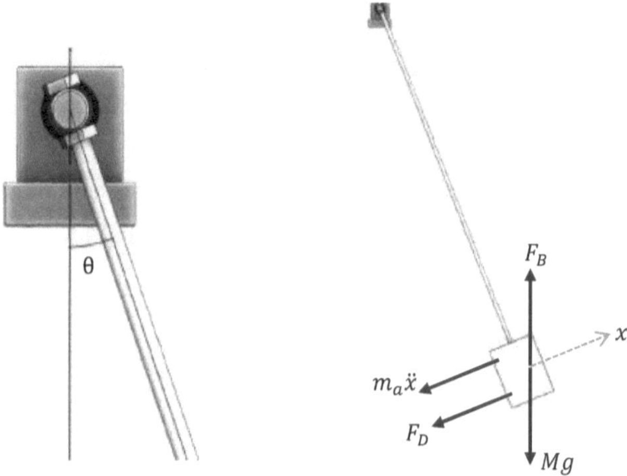

Figure 50: Forces on block at angle of lift θ

Assuming the rod is of a negligible weight, the total Newtonian forces acting on the part can be summed up as follows [3]:

$$\sum F_x = M\ddot{x} \qquad \{6.1\}$$

$$F_B \sin\theta - Mg \sin\theta - m_a\ddot{x} - F_D = M\ddot{x} \qquad \{6.2\}$$

$\ddot{x}$ is the linear acceleration, F_B is the buoyancy force, Mg is the part's dry weight, F_D is the damping force and $m_a\ddot{x}$ is the added force due to the surrounding fluid's induced unsteady motion. Taking into account the linear effects of skin friction, drag force directly proportional to velocity was included in the model. Likewise, a drag force directly proportional to velocity squared represents the effects of vortex shedding (pressure drag) and the non-linear skin friction on the submerged part. Equation 6.2 can therefore be expanded as follows,

$$(F_B - Mg)\sin\theta - m_a\ddot{x} - K_L\dot{x} - K_Q\dot{x}|\dot{x}| = M\ddot{x} \qquad \{6.3\}$$

Where K_L is the linear damping coefficient and K_Q is the quadratic damping coefficient. The damping coefficients were assumed to be constant although in reality are a function of the fluid's compressibility and viscosity. Higher order damping were assumed to be negligible. Rearranging the equation while also writing it in rotation motion form,

$$\ddot{\theta} = \frac{F_B - Mg}{(M+m_a)r}\sin\theta - \frac{K_L}{M+m_a}\dot{\theta} - \frac{K_Q r}{M+m_a}|\dot{\theta}|\dot{\theta} \qquad \{6.4\}$$

To simplify the equation, the following assignments were made [3]:

$$\alpha = \frac{F_B - Mg}{(M+m_a)r}, \quad \beta = \frac{-K_L}{M+m_a}, \quad \gamma = \frac{-K_Q r}{M+m_a} \qquad \{6.5\}$$

Therefore, equation 6.4 becomes,

$$\ddot{\theta} = \alpha\sin\theta + \beta\dot{\theta} + \gamma|\dot{\theta}|\dot{\theta} \qquad \{6.6\}$$

As the $\theta, \dot{\theta}$ and $\ddot{\theta}$ values are obtainable experimentally, the equation can be vectorised to contain a number of rows depending on the number of samples taken [3],

$$\begin{bmatrix} \ddot{\theta}_1 \\ \ddot{\theta}_2 \\ \vdots \end{bmatrix} = \begin{bmatrix} \sin\theta_1 & \dot{\theta}_1 & |\dot{\theta}_1|\dot{\theta}_1 \\ \sin\theta_2 & \dot{\theta}_2 & |\dot{\theta}_2|\dot{\theta}_2 \\ \vdots & \vdots & \vdots \end{bmatrix} \begin{bmatrix} \alpha \\ \beta \\ \gamma \end{bmatrix} \qquad \{6.7\}$$

This system of equations could then be rewritten as,

$$Y = Xb \qquad \{6.8\}$$

Using the method of *least squares regression*, this equation can then solved for b, and ultimately m_a can be obtained by rearranging the equation with α from equation 6.5.

$$b = (X^T X)^{-1} X^T Y \qquad \{6.9\}$$

As stated previously, the values of $\theta, \dot{\theta}$ and $\ddot{\theta}$ will be experimentally determined, depending on the sensors installed on the oscillating component. The values $\dot{\theta}$ and

θ could also be obtained by differentiating the response twice, using software such as *Excel* or MATLAB's *Curve Fitting Toolbox* applications.

6.1.2 Experimental Method Verification

Two different methods were used to check the feasibility of the suggested experimental approach: by performing a pendulum arc length check and by performing a translational added mass check. The tests were performed on three components. The reason for this will be subsequently stated.

<u>Method 1</u>

This involved comparing the experimentally estimated oscillation arc length to its measured counterpart for each component. The oscillation was performed in the air with negligible drag, therefore having very low decay rate. This was performed on all three components. The least square computation previously explained was used for the estimation. Rather than solving for m_a (since it is close to zero in the air), r was determined.

<u>Method 2</u>

This method compared the theoretical added mass of each of the three parts to its equivalent experimental estimate. The experimental added masses were obtained by performing a free oscillation in the water tank and using the recorded response data in a least square method MATLAB algorithm. The purpose of doing this (i.e. performing the same experiment on all three parts – one cuboid, two smaller versions of the cuboid but of different densities) was to verify that the mass property of a part does not affect the added mass, but rather the main factor is the part's geometry. It also served to verify the scalability of added mass from the comparison of the added mass of the simple cuboid to those of the 2 smaller parts.

Once the verification stage had been completed, the same test would be carried out on a small-scale model of the Seahog. The setup for all ensuing testing was such that the component was attached to the end of the pendulum rod which is joined to the connecting shaft which has a calibrated potentiometer attached to its other end. The potentiometer used was a series 6539 Bourns precision potentiometer [71] whose angular position was sampled at a rate of 100 Hz by an STM32F0 microprocessor based development board. The pendulum and potentiometer were connected as shown in Figure 51.

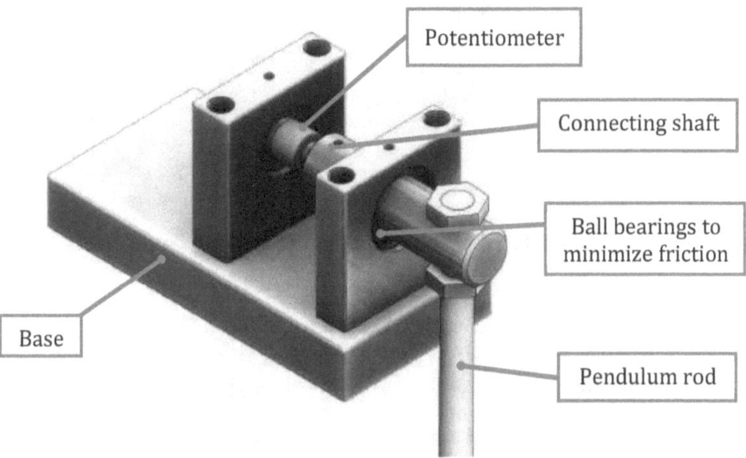

Figure 51: Pendulum - potentiometer setup

Angular position and their corresponding time details from the saved pendulum motion were then exported and used in a MATLAB function that basically outputs an added mass value, taking in the aforementioned quantities as inputs. The MATLAB function created for this purpose was based on the Newtonian composition described in the previous section.

The three parts made for the verification step were as follows: one simple cuboid and two smaller cuboids of similar size but different densities in order for them to possess different mass properties. The size ratio of the big cuboid to the small cuboids was two as shown in the following figure.

Figure 52: Parts made for experiment verification from left to right: main cuboid made from PVC, small scaled cuboid made from PVC and small scaled cuboid made from aluminium.

The following table details the properties of each part.

Table 22: Test component parts and their characteristics

Part	Material	Dimension (mm × mm × mm)	Mass
Simple cuboid	PVC	100 x 100 x 130	1884 g
Small scaled cuboid 1 (SSC PVC)	PVC	50 x 50 x 65	235 g
Small scaled cuboid 2 (SSC Al)	Aluminium	50 x 50 x 65	439 g

6.1.2.1 Theoretical Added Mass

In this section, the main focus will be on the approach taken in obtaining the aforementioned theoretical added mass of the simple PVC cuboid block of dimension 100 x 100 x 130 mm. The same approach was then applied to the other cuboid components. Added mass coefficients of rectangular cross-section cylinders were obtained from the technical maritime recommended practices document compiled by DNV [52] and the report by Chung and Chen [56]. In these documents, an added mass coefficient C_A was given for a corresponding ratio b/a, where b and a represent the rectangular cross-section lengths. Using Excel's best fit function, a curve of C_A vs b/a was generated with the given points.

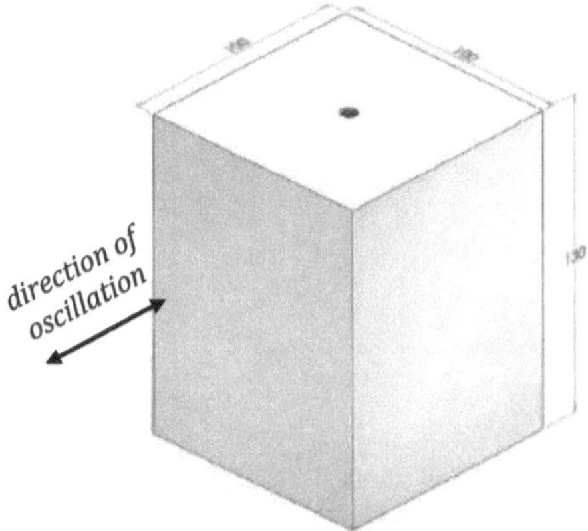

Figure 53: Direction of cuboid's oscillation

When working with added mass coefficients, note that if the cross-sectional area of a cuboid perpendicular to the direction of oscillation is of the form: $x \times y$, where $x \neq y$, it is then valid to take the effective length a [72] as:

$$a = \sqrt{x \times y} \qquad\qquad \{6.10\}$$

The following is a graph showing the relationship between the b/a ratio and the added mass coefficient C_A. Note that the double-sided arrow beside the part displayed on the graph represents the direction of motion of the part for which the coefficients are valid.

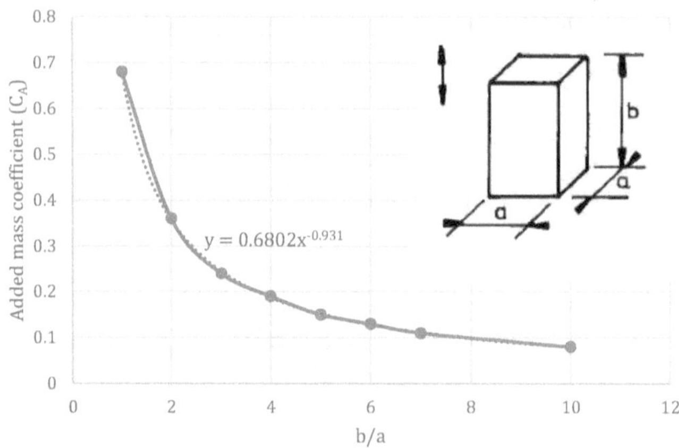

Figure 54: Added mass coefficient vs b/a ratio for the cuboid

It should be noted that the graph above displays the range of values provided by DNV [52], but in actual fact, there is experimental data provided by Chung and Chen [56] extending this table up to b/a ratio of 0.5. The best of fit's equation accurately predicts the added mass coefficient to this point. The added masses were obtained using the following formula [52], where V_R is the reference volume:

$$m_a = \rho C_A V_R = \rho C_A a^2 b \qquad \{6.11\}$$

The added masses of the scaled models (the small PVC and aluminium parts) can be obtained by dividing the cuboid's added mass by eight as the scaled model is half the size of the simple cuboid, therefore having a volumetric scale factor of 2^3.

Table 23: Theoretical added masses

Part	b/a	C_A	m_a
Simple cuboid	$100/\sqrt{100 \times 130} = 0.877$	0.769	1.000 kg
SSC PVC	0.877	0.769	0.125 kg
SSC Al	0.877	0.769	0.125 kg

6.1.2.2 Experimental Added Mass

The tests were carried out in a water tank of size 1 m x 3 m x 1 m, filled with fresh water to a depth of 0.5 m. This was deemed a sufficient testing volume for this particular experiment.

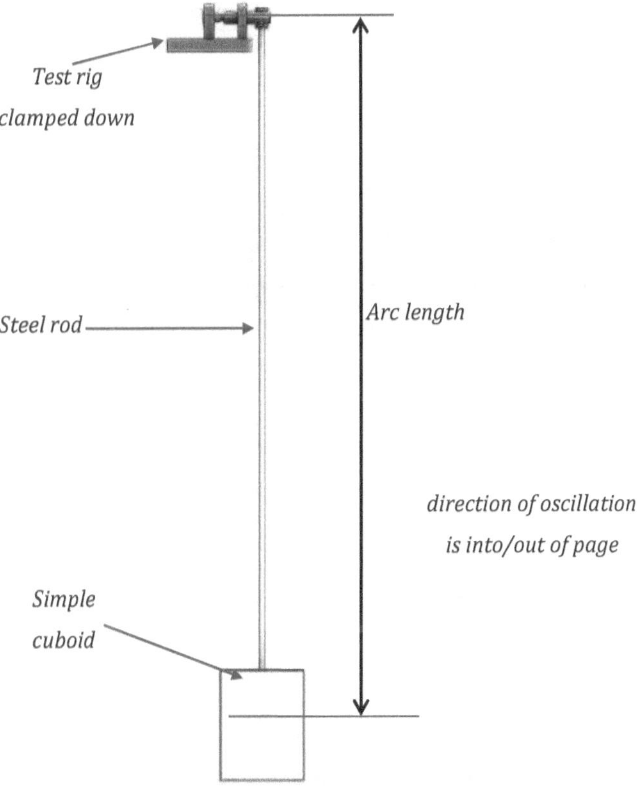

Figure 55: Experimental system setup

The following points summarize the standard pendulum test procedure followed:

- The test rig was connected to the relevant component through the pendulum rod.

- The part was oriented such that the relevant cross-section was set perpendicular to the direction of motion.

- The rod was lifted to a suitable angle and released to freely oscillate.

- The oscillation's trajectory was tracked by a potentiometer and saved to an Excel file through a microcontroller at a specified sampling rate. This sampling rate took into account the minimization of noise on the pin being used to sample the data.

- The recorded trajectory was then exported to and plotted on MATLAB and the added mass was obtained through curve fitting and least squares algorithm explained in section 5.2.1.

- This procedure was repeated three times for each component.

The next figure is a typical oscillatory response obtained from performing the experiment. This was taken from one of the data samples recorded.

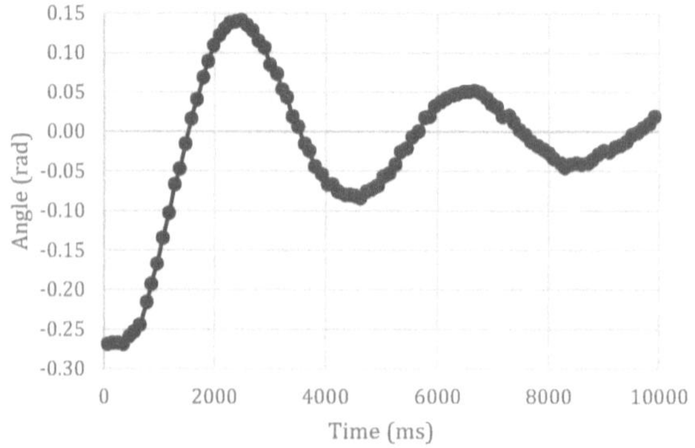

Figure 56: An example of a typical response obtained from the experiment

6.1.2.3 Verification Process Results

<u>Method 1</u>

The results from the tests can be seen on Table 24. It also includes the standard deviation for each set from the mean value. The experiment was performed three times on each cuboid. Viewing each experiment performance as a sample point, i.e.

three times on a component equals three sample points, this imposed a limitation on the calculated standard deviation as the sole qualifier of the experimental method result's precision level. This is because a minimum of 30 sample points is typically required when describing the standard deviation of a parameter. The *margin of error* associated with a set's mean result was therefore also considered. Equation 6.12 shows how the margin of error is typically calculated using the sample's standard deviation s of samples of size n and a confidence coefficient κ.

$$margin\ of\ error = \kappa\frac{s}{\sqrt{n}}, \qquad where\ \kappa = 4.303 \qquad \{6.12\}$$

κ which was a function of the t-distribution was based on an assumption of a 95% chance of the true value occurring within the *confidence interval* (mean ± margin of error) if the experiments were repeated. This was also applied to method 2.

Table 24: Experimental arc length results from method verification

Oscillation arc length (m)

	Test no.				Sample's	Margin
	1	*2*	*3*	*Mean*	*std. dev.*	*of error*
Simple cuboid	0.6741	0.6733	0.6769	0.675	0.0019	0.0047
SSC PVC	0.5842	0.5984	0.5783	0.587	0.0103	0.0257
SSC Al	0.6298	0.6324	0.6217	0.628	0.0056	0.0139

<u>Method 2</u>

The following table details the results of the method 2 tests.

Table 25: Experimental added mass results from method verification

Added mass (kg)

	Test no.				Sample's	Margin
	1	*2*	*3*	*mean*	*std. dev.*	*of error*

Simple cuboid	1.2655	1.0629	1.1993	1.176	0.1033	0.2567
SSC PVC	0.1198	0.1126	0.1114	0.115	0.0045	0.0113
SSC Al	0.1138	0.1389	0.1083	0.120	0.0163	0.0405

6.1.2.4 Result Analysis

The two previous tables provided a detailed report on the results from both verification methods. The added mass and arc length estimates needs to be compared to their respective ideal values to gauge accuracy, i.e. the closeness of the estimated values to their real or theoretical counterparts. The following two tables serve this purpose. The percentage differences stated were based on the standard deviation from the ideal value for each cuboid. On the other hand, Table 24 and Table 25 were made for analysing the experiment's precision, i.e. the measure of the experiment's repeatability. This was why the standard deviations and margins of error stated were computed independent of the ideal arc lengths and added masses.

Table 26: Measured and experimentally obtained arc length comparison

	Measured arc length	*Average experimental arc length*	*% difference*
Simple cuboid	0.673 m	0.675 m	0.35%
SSC PVC	0.643 m	0.587 m	8.81%
SSC Al	0.643 m	0.628 m	2.44%

The results from both the theoretical analysis and the experiment verification method 2 are compared in the following table.

Table 27: Theoretical and the mean of the experimentally obtained added mass comparison

	Theoretical	Experimental mean	% difference
Simple cuboid	1.000 kg	1.176 kg	19.51%
SSC PVC	0.125 kg	0.115 kg	8.83%
SSC Al	0.125 kg	0.120 kg	11.29%

From Table 26, the 'heavier' cuboids had smaller deviations from their respective ideals than the small PVC while the experiment was run in the air. While executed in water to determine added mass, the small PVC had the smallest deviation, though not significantly smaller than the small aluminium cuboid's. Table 27 showed a maximum added mass deviation from the theoretical value to be around 19%. These and the other discrepancies could be attributed to the empirical added mass coefficients used in obtaining the theoretical values. These values were obtained experimentally, and like most experiments, there had to be some uncertainties associated with the process and therefore, the ensuing added mass coefficients. There were factors in the pendulum experiment itself that could have added more error sources to the process. Some of these factors will be discussed at the end of this chapter.

From these results, the free oscillation method has been shown to be an approach with outputs that were relatively accurate when compared to their expected theoretical values. Though the small cuboids were of different masses, their % differences of about 9% and 11% respectively were seen as relatively small. This therefore demonstrated that added mass is indeed independent of mass but rather shape and volume. The method's precision level which is based on the margins of error recorded (Table 24 and Table 25) were also considered to be adequate as the ideal added masses were observed to fall within the confidence intervals for all the cuboids. This was not the case for the arc lengths as only the PVC cuboid had the ideal arc length fall within its confidence interval. This was to be expected since a

95% confidence level implies there exists a 5% chance that an experiment would not result in an interval with the ideal value. The pendulum method was therefore a viable approach for estimating the Seahog's translational added masses through a scalable small-scaled model. The next section describes the design process undertaken in making a scaled model of the Seahog.

6.1.3 Mini-ROV Design

To be able to perform the pendulum test under the most ideal conditions, it was deemed necessary to make a small-scaled model of the Seahog. This was due to the limiting size of the available water tank. The design of the miniaturised version of the Seahog had to match the overall geometry of the Seahog as fluid interacts with the surface boundary of the Seahog. The scaling factor to be employed had to give critical consideration to the test apparatus' loading and measuring limits, and the water tank's size. In order to try to maintain the theoretical assumption of infinite fluid body as much as possible, thereby minimising the shallow or narrow water effect on added mass, a minimum size factor ratio (with regards to the water depth/width to the vehicle's diametric length ratio) of 10 would be ideal. The factor of 10 was specified by Zhou et al. [65] who found the added masses of three cylinders of varying cross-sections being numerically analysed tended to be fully converged at around a domain size of approximately 10 or even smaller. 10 was the domain size used when treating the water body as wide or deep. After all these factors were taken into account, a scale factor of four was applied to the Seahog to produce a small scaled version (generally referred to as mini-ROV in this document). Figure 57 shows the overall look and size of the mini-ROV. Considering the mini-ROV positioned in the surge direction along the tank's longitudinal length, the water tank to mini-ROV size ratios were slightly less than 10. This was deemed suitable for the experiment as a smaller mini-ROV made from the same material might not have provided sufficient oscillatory response due to drag effects.

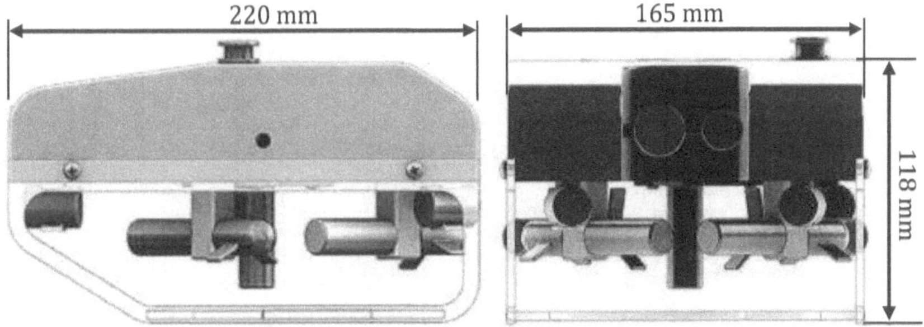

Figure 57: Mini-ROV's overall size

Simplifications were made to the design were due to manufacturing and functional considerations such as the time taken to make the redesigned propellers was shorter than it would have taken to make the exact replica of the Seahog's due to its complex geometry. The Seahog has space in which water can flow between its cover and its top hull, but this was not included in the small scale design. This was deemed appropriate since the water flowing within that cavity can be considered to be a part of the Seahog's weight. The following figures display some other differences in design between the Seahog (left) and the mini-ROV (right). The hull (top part) and the thruster parts of the mini-ROV were made from aluminium, while the bottom frame was made from Perspex.

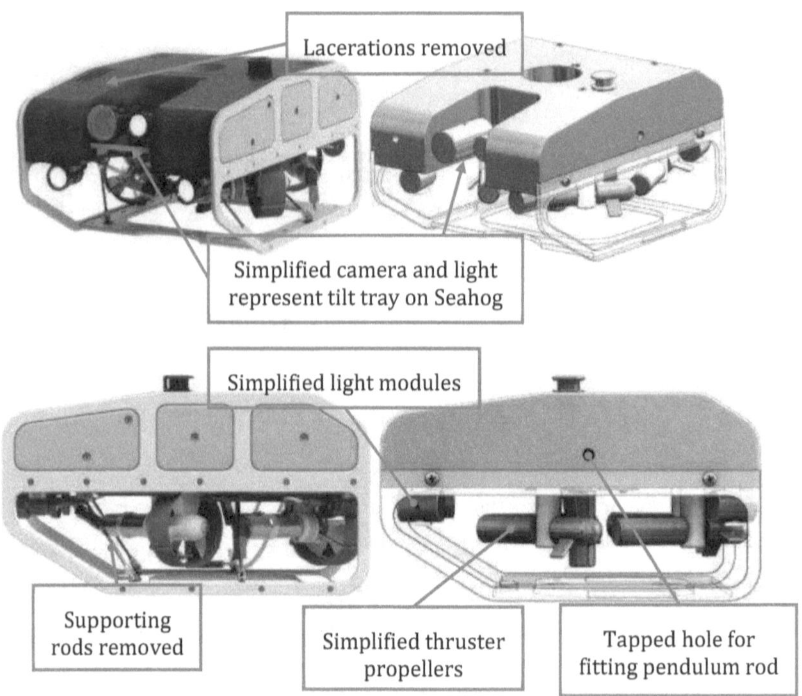

Figure 58: Mini-ROV vs Seahog model

6.2 Results

The mini-ROV's added mass values obtained from the each set of experimental data analysis were as follows.

Table 28: Results from the least square computations using the experimental data

	Added mass (kg)		
Instance	Surge	Sway	Heave
1	1.017	2.554	4.448
2	1.180	2.145	5.294

3	1.314	2.569	5.156
Average	1.170	2.423	4.966

The added mass coefficients C_a were obtained using the relationship between added mass m_a, fluid density (freshwater in this case) ρ and volume V of the mini-ROV:

$$m_a = C_a \rho V \rightarrow C_a = \frac{m_a}{\rho V} \qquad \{6.13\}$$

Using the added mass coefficient, the Seahog's translational added masses were then obtained and shown on Table 29.

Table 29: Seahog's added mass along each translational direction

Added mass term	Added mass coefficient	Equivalent Seahog added mass
Surge	0.68	47.85 kg
Sway	1.41	99.05 kg
Heave	2.89	203.04 kg

The following table documents the new added mass values with their corresponding prior values.

Table 30: New translational added masses

Direction	New added mass	Current added mass
Surge	47.85 kg	42.1 kg
Sway	99.05 kg	57.2 kg
Heave	203.04 kg	359.9 kg

6.3 Added Mass Implementation in the Simulation

The new added mass values were subsequently implemented in the Seahog's Simulink-based simulation. The following are two sets of results from two Seahog motion simulation. In the first simulation, a depth set point was set for the Seahog. The chart on the left shows the Seahog's localization response with the previous translational added masses implemented while the chart on the right shows the response to the new added masses.

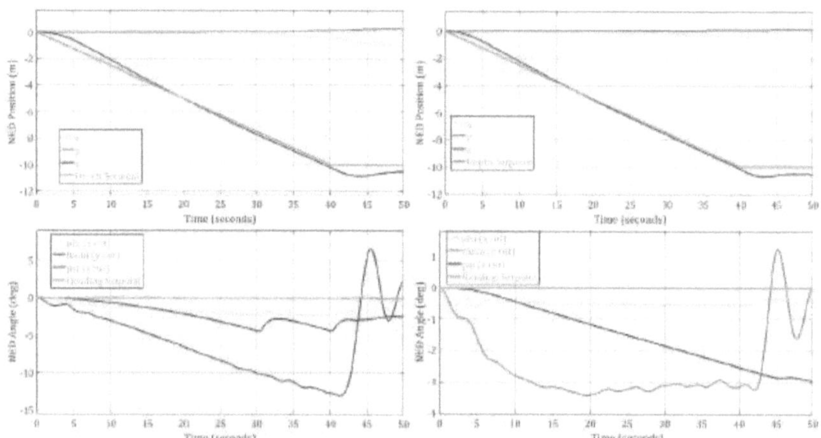

Figure 59: Simulation response with old added mass (left), simulation with new added mass (right)

The second sequence was a surge motion followed by a heading motion. The following graphs follow the order of arrangement as the previous figure.

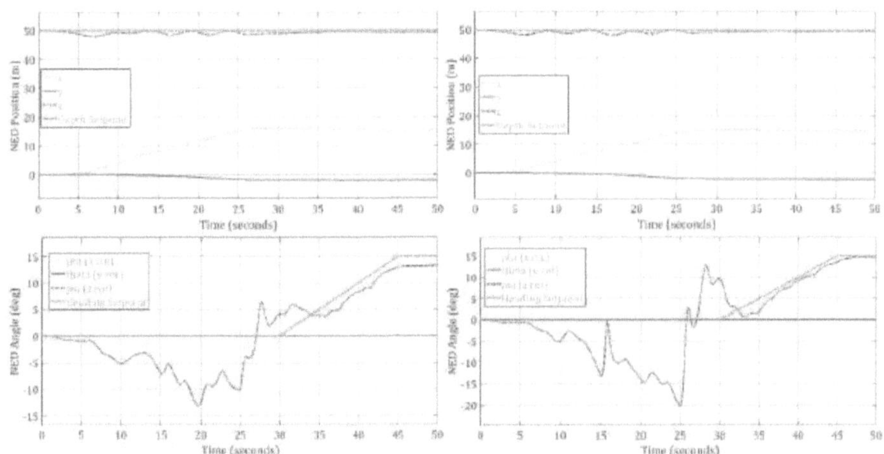

Figure 60: Simulation response with old added mass (left), simulation with new added mass (right)

Note that the simulation contains a random number generator used in implementing the depth sensor noise model. This implies that for such variable, random numbers were generated for every new iteration of the simulation.

6.4 Conclusion

In comparison to the previous Seahog model used for estimating the translational added mass, the mini-ROV was a more accurate representation of the Seahog. The testing methodology had also been shown to produce fairly accurate and precise results. The added mass results obtained could therefore be taken to be more accurate than the previous values. This was also validated by the fact that the new values were as initially predicted in section 5.2.3, where the new surge and sway values are greater than their previous counterparts and the new heave is less than the previous value.

There were however some factors in the methodology of the test that could have affected the accuracy of the final results. These factors will be discussed in the next section.

6.4.1 Shortcomings of the Experiment

As seen in the experiment's outputs, variations existed between values
obtained under the same test conditions. During the method verification phase,
there were also variations between the theoretical and experimental added mass
values. Some
of the factors that could account for these discrepancies are as follows:

- The theoretical equations assume an infinite volume of water i.e. the body is
  far away from any boundary [73], but the experiment was carried out in a
  confined tank which was assumed to be sufficient for this project. However,
  as the water tank being of size 1 m x 3 m x 1 m can be considered to be quite
  narrow, meaning that water flow reverberation could have had a significant
  effect on the results.

- As seen in the oscillatory response graph Figure 56, in addition to vibratory
  noise from the pendulum's rod, the disturbances occurring along some parts
  of the oscillation could be due to electrical noise along the connection
  between the potentiometer and the data processing microprocessor.

- The added mass coefficients were assumed to be constant for fully
  submerged vehicles, thus independent of the effects of depth travelled in the
  fluid, proximity to boundaries and the body's relative acceleration.

- Differences in repeated experiment results might be due to slight differences
  in initial angular positioning of the pendulum during testing. It was crucial
  for the initial positioning of the pendulum to be constant for repeated tests
  but slight differences (typically < 10%) were observed.

- The free decay model ignores the losses due to the frictional losses in the
  pendulum rod and potentiometer's bearings.

- The pendulum rod's weight was neglected in the calculations but in reality,
  could have had some impact, though insignificant on some of the results. The
  weight might have been especially significant for the small PVC cuboid due
  to its small weight as its oscillation was a lot more damped than the other
  cuboids, thereby amplifying the various noise signals.

6.4.2 Potential System Improvement

Even though the translational added masses have been successfully identified, further work still needs to be done to better the results or verify the other crucial hydrodynamic parameters that further defines the Seahog-water interaction.

- Firstly, a system identification test on the mini-ROV could be performed in a suitably sized basin. The basin could be of minimum size 2.2 m × 1.6 m × 1.2 m, although a basin with a longer length than the specified minimum would be preferable to accommodate for translational motion along that length.

- The significance of depth travelled on added mass can be studied. It was not possible here because the water tank used was not deep enough for a range of varying pendulum lengths to be tested. This is especially pertinent to the Seahog's operation in shallow waters or near-surface operations according to research findings by Zhou et al. [65] and the DNV [52] guidelines.

- The hydrodynamic effects on the Seahog due to the taut tether attached to it could be studied. This would serve in the overall improvement of the simulation model and further understanding of its dynamics.

- An experimental or theoretical formulation could be implemented to identify more accurate rotational added moments of inertia values.

- Once the Seahog has been fully realised mechanically, a full system identification exercise including the identification of all hydrodynamics forces could be performed in a tow tank or equivalent facility with PMM facilities to help improve its simulation model.

7 PROJECT SUMMARY AND RECOMMENDATIONS

This chapter serves to summarise the work done in this project and recommend potential areas of improvement. The following is a recap of the objectives set out at the beginning of this book:

1. Design and test an attitude estimating subsystem to be implemented on the Seahog.
2. Design, test and integrate the tilt unit controller with the front camera and light system.
3. Investigate the hydrodynamic added mass model of the Seahog used in the simulation and control system model.

Chapters 2 – 4 detailed the work done in achieving objective 1. Chapters 2 and 3 were a form of introductory literature while chapter 4 detailed the actual work breakdown, execution and analysis. A system that could estimate the attitude of the Seahog was implemented on the Seahog's inertial sensor's embedded processor.

Chapters 5 and 6 document the work done in added mass estimation (objective 3). These chapters also followed the same format as the previous ones; chapter 5 covered literature review while 6 detailed the work done in obtaining added mass values. The following is a table detailing the new added mass values implemented in the simulation in comparison with the old values.

Table 31: New vs old added mass values

Direction	New added mass and inertia	Old added mass and inertia
Surge	47.85 kg	42.1 kg
Sway	99.05 kg	57.2 kg

Direction	New added mass and inertia	Old added mass and inertia
Heave	203.04 kg	359.9 kg
Roll	7.0 kgm²	7.0 kgm²
Pitch	16.0 kgm²	16.0 kgm²
Yaw	4.1 kgm²	4.1 kgm²

To fulfil objective 2, a magnetic coupling-based tilt unit controller was designed and tested successfully. As stated earlier in chapter 1, the bulk of the groundwork required to meet objective 2 had already been covered and documented by Hope [13]. For this reason, the report on the process of fulfilling this objective was placed under the Appendices section and can be found in Appendix A.

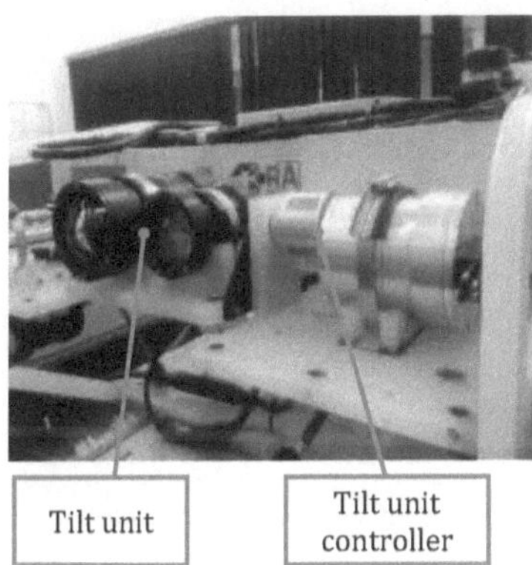

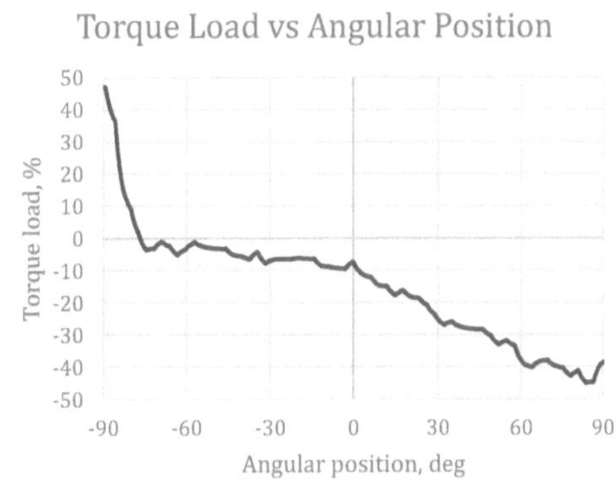

Figure 61: Left - tilt unit controller connected to tilt tray, right - torque load at various tilt unit angular positions

Figure 61 shows the tilt controller as it was installed on the Seahog and the torque required by the tilt unit over the angular position span of 180 °, taking the positive angle as upwards facing, with respect to the Seahog's positive x-axis.

7.1 Further Seahog Improvements

At the end of chapters 4, 6 and Appendix A, additional tasks were set out that could be executed to improve attitude estimation, added mass characterization and tilt unit controller performance respectively. To get the Seahog to a state beyond basic functionality however, more work could be carried out. The potential work to be carried out in the future have been divided into two categories: further improvements on gains made during this project and general tasks unrelated to this project.

7.1.1 Potential Project Improvement

- **Localization**: Though the attitude system developed for the Seahog was relatively successful in meeting the performance target, particularly the roll and pitch estimates, the yaw was ultimately based on angular rate integration. Implementing this on the Seahog as its first attitude estimation system is fine in a controlled or fairly 'static' environment. The non-transient pitch and roll estimates produced would be accurate to within a 1 ° margin. This needs to be verified through field testing. The yaw estimator's performance on the other hand is expected to deteriorate over time and inaccurate in an unstable environment. With the average ocean temperatures along the Western Cape coastline being in the region of 12 – 20 °C, the accelerometer and gyroscope's performances could be investigated at low temperatures. This could be achieved by performing a field test in waters or an environment within the typical coastal ocean temperature range. Sensor readings could also be taken at high temperatures to simulate the situation where the iNEMO's module is experiencing high temperatures from cases such as ineffective heat dissipation. Another strong suggestion is that a dedicated AHRS unit be obtained for the Seahog. It was suggested at the end of chapter 4 that a gyrocompass be obtained for better yaw estimation but this would result in the attitude system requiring multiple hardware. This could potentially demand additional power from the Seahog's power supply and add a communication delay that could be avoided with a dedicated AHRS unit. With regards to the use of the magnetometer, there are yaw estimation algorithms that include the magnetic field measurements which compensate for magnetic noise. Some of these work by compensating for a pre-defined magnetic signature while others are more robust. Further work could be carried out in investigating the feasibility of implementing such an algorithm. Works by Fan et al. [35] and Yadav and Bleakley [36] serves as examples. Installation of a velocity sensor would aid in the positional localization of the Seahog which is also crucial to navigation.

- **Added Mass and Inertia Characterization**: Only the translational added masses were obtainable through the experimental method employed. The easiest way of obtaining reliable translational and rotational added masses is computationally. As stated in section 5.1, computational packages dedicated for this purpose are expensive and was well beyond this project's budget. Except a CFD package capable of calculating added masses is obtainable, work could be done in finding alternative approaches to solve this problem, particularly for solving for the rotational added masses (moment of inertia). The added mass values could also be directly verified experimentally in a controlled environment. The effects of depth and relative fluid motion on added mass could also be investigated.

- **Tilt Unit Angular Positioning Control**: The outer to inner rotor connection could be redesigned to ensure accurate alignment with the tilt unit tray's rotation axis. This problem was unfortunately neglected in the original design. A fairly accurate alignment could be achieved by redesigning the tilt controller's motor housing and its polypropylene seat to include a locating feature such as a groove and tongue. Alignment is currently achieved manually and heavily relies on the aluminium cable tie to maintain it (see Figure 80).

7.1.2 General Improvement of the Seahog

- **Improve on the Seahog's Simulation Model:** This can be achieved by incorporating environmental factors like fluid flow turbulence. Incorporating the tether dynamics into the simulation model would also be recommended. The tether is expected to have an effect on the Seahog's dynamics and kinematics due to its weight properties and how that affects the Seahog's overall dynamics.

- **Design or Procure a Tether Management System:** Currently, the Seahog's tether is wound around a wooden spool which needs to be manually wound by an operator to release and retract the tether during operations.

- **Redesign the Junction Box:** The junction box is the Seahog's module connected to the tether. It is through it that surface power and communication systems are linked to the Seahog. The current oil-filled junction box-to-tether connection is intricate and difficult to assemble and disassemble. It should be noted though that this is not an urgent issue and is not expected to affect the Seahog's performance.